Florian Ion **PETRESCU** &
Relly Victoria **PETRESCU**

CAMSHAFT

PRECISION

-COLOR-

2012

Scientific reviewer:

Dr. Veturia CHIROIU

Honorific member of
Technical Sciences Academy of Romania (ASTR)
PhD supervisor in Mechanical Engineering

ISBN 978-1-4802-9103-4

WELCOME

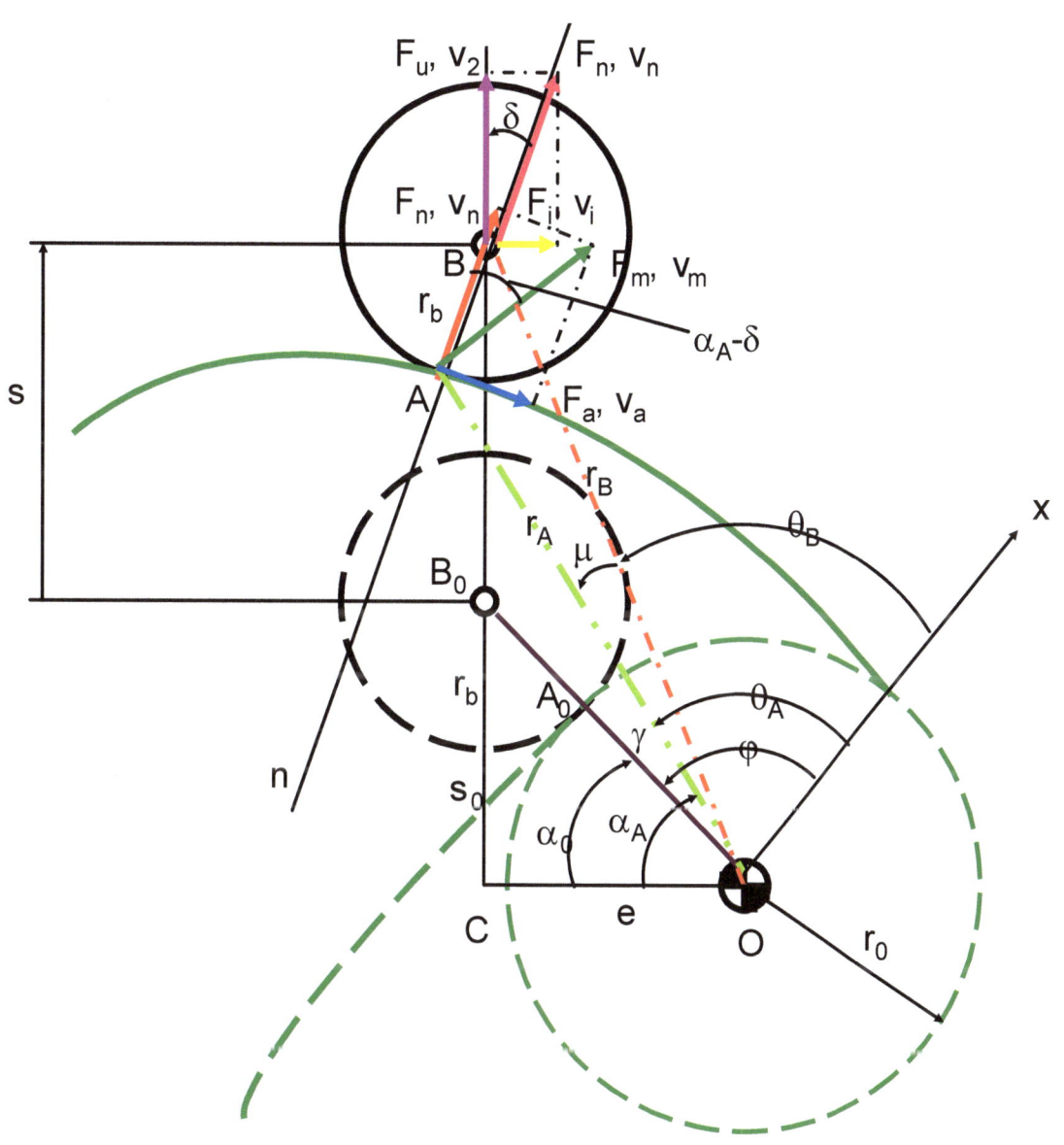

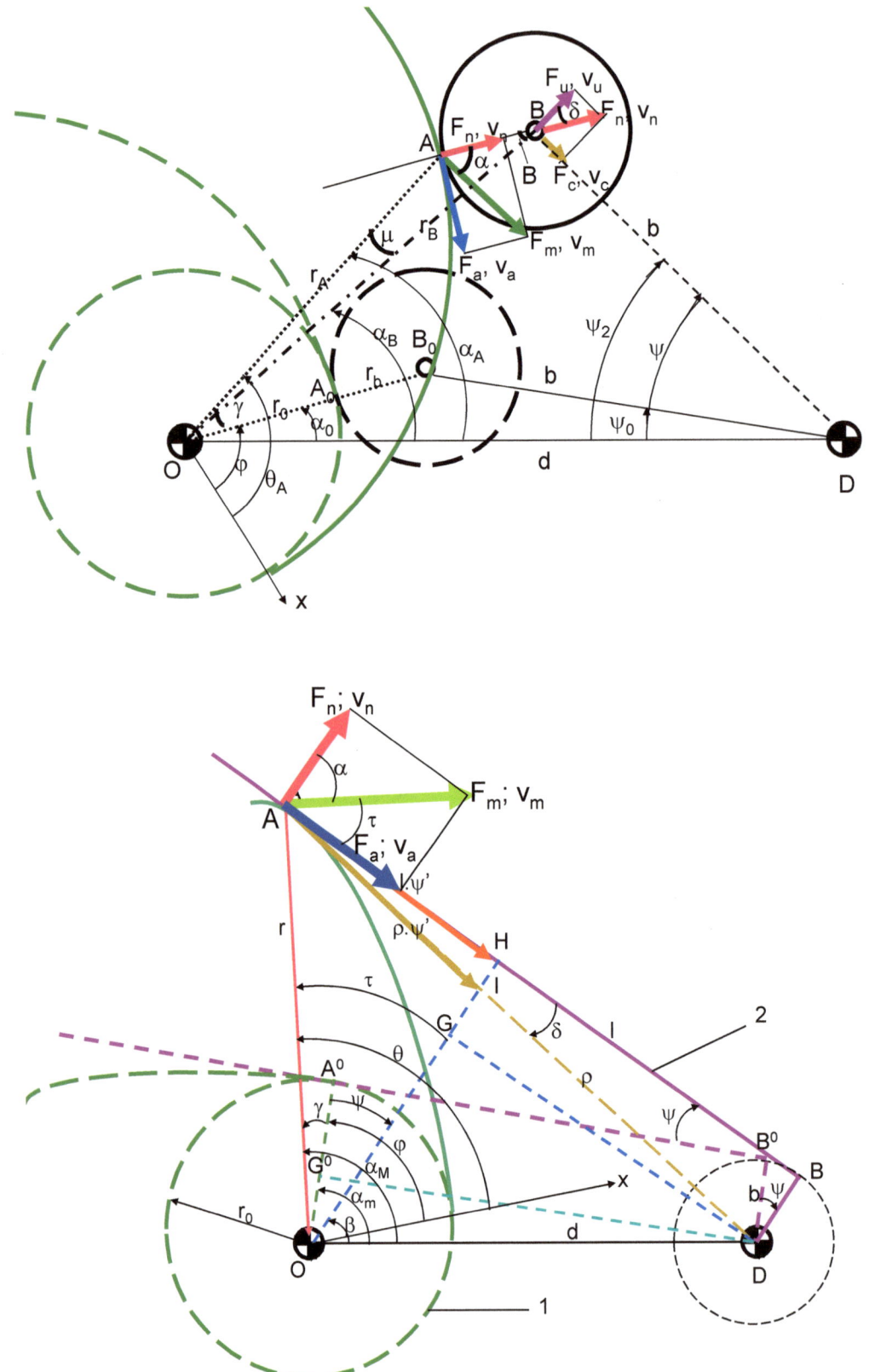

You are welcome to read the full book! The authors.

CONTENT

Welcome.. 003

Content... 005

Cap 01 CAM GEARS EFFICIENCY.......................... 006

Cap 02 CONTRIBUTIONS AT THE
DYNAMIC OF CAMS 014

Cap 03 CAM GEARS DYNAMICS ILLUSTRATED
IN THE CLASSIC DISTRIBUTION 021

Cap 04 CAM GEARS DYNAMICS
TO THE MODULE B (WITH TRANSLATED
FOLLOWER WITH ROLL).................................. 031

Cap 05 DYNAMICS OF THE
CLASSIC DISTRIBUTION.................................. 042

Cap 06 PRECISION OF THE
CLASSIC DISTRIBUTION.................................. 048

Cap 07 DYNAMIC SYNTHESIS OF THE ROTARY
CAM AND TRANSLATED TAPPET WITH ROLL.......... 058

Bibliography... 070

Annex... 077

CHAPTER I

CAM GEARS EFFICIENCY

Abstract: *The chapter presents an original method to determine the efficiency of a mechanism with cam and follower. The originality of this method consists in eliminating the friction modulus. In this chapter it analyses four types of cam mechanisms: 1.The mechanism with rotary cam and plate translated follower; 2.The mechanism with rotary cam and translated follower with roll; 3.The mechanism with rotary cam and rocking-follower with roll; 4.The mechanism with rotary cam and plate rocking-follower. For every kind of cam and follower mechanism one uses a different method to determine the most efficient design. We take into account the cam's mechanism (distribution mechanism), which is the second mechanism in internal-combustion engines. The optimizing of this mechanism (the distribution mechanism), can improve the functionality of the engine and may increase the comfort of the vehicle too.*

Keywords: *cam, efficiency, translated follower, rocking-follower, follower with roll*

1 Introduction

In this chapter the authors present an original method to calculate the efficiency of the cam's mechanisms. Four kinds of cam and follower mechanisms are analyzed: 1. A mechanism with rotary cam and plate translated follower; 2. A mechanism with rotary cam and translated follower with roll; 3. A mechanism with rotary cam and rocking-follower with roll; 4. A mechanism with rotary cam and plate rocking follower. For every kind of cams and followers mechanism, a different method for the cam's design with a better efficiency has been utilized.

2 Determining of momentary mechanical efficiency of the rotary cam and plate translated follower

The consumed motor force, F_c, perpendicular at A to the vector r_A, is divided into two components [1, 2]: a) F_m, which represents the useful force, or the motor force reduced to the follower; b) F_ψ, which is the sliding force between the two profiles of cam and follower (Fig. 1). See the written relations (2.1-2.10):

$$F_m = F_c \cdot \sin \tau \qquad (2.1)$$

$$v_2 = v_1 \cdot \sin \tau \qquad (2.2)$$

$$P_u = F_m \cdot v_2 = F_c \cdot v_1 \cdot \sin^2 \tau \qquad (2.3)$$

$$P_c = F_c \cdot v_1 \qquad (2.4)$$

$$\eta_i = \frac{P_u}{P_c} = \frac{F_c \cdot v_1 \cdot \sin^2 \tau}{F_c \cdot v_1} = \sin^2 \tau = \cos^2 \delta \qquad (2.5)$$

$$\sin^2 \tau = \frac{s'^2}{r_A^2} = \frac{s'^2}{(r_0 + s)^2 + s'^2} \qquad (2.6)$$

6

$$F_\psi = F_c \cdot \cos \tau \qquad (2.7)$$

$$v_{12} = v_1 \cdot \cos \tau \qquad (2.8)$$

$$P_\psi = F_\psi \cdot v_{12} = F_c \cdot v_1 \cdot \cos^2 \tau \qquad (2.9)$$

$$\psi_i = \frac{P_\psi}{P_c} = \frac{F_c \cdot v_1 \cdot \cos^2 \tau}{F_c \cdot v_1} = \cos^2 \tau = \sin^2 \delta \qquad (2.10)$$

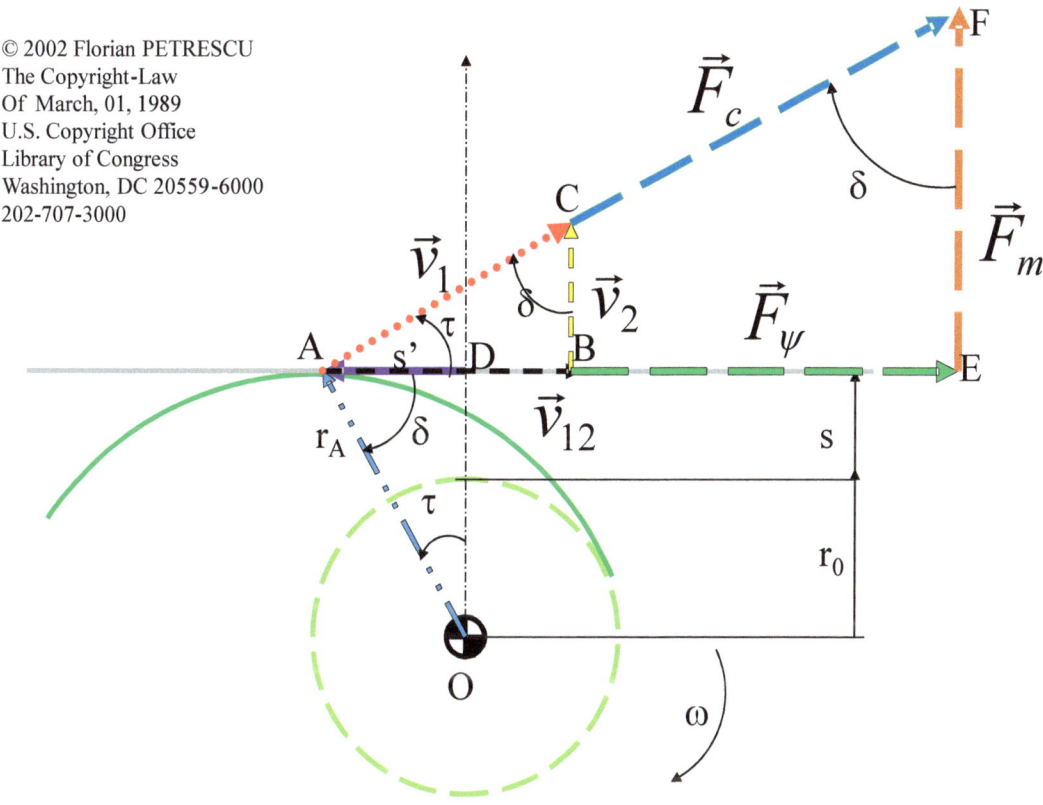

Fig. 1 *Forces and speeds to the cam with plate translated follower*

3 Determining of momentary dynamic efficiency of the rotary cam and translated follower with roll

The pressure angle δ (Fig. 2), is determined by relations (3.5-3.6) [1, 2]. We can write the next forces, speeds and powers (3.13-3.18). F_m, v_m, are perpendicular to the vector r_A at A. F_m is divided into F_a (the sliding force) and F_n (the normal force). F_n is divided too, into F_i (the bending force) and F_u (the useful force). The momentary dynamic efficiency can be obtained from relation (3.18):

The written relations are the following.

$$r_B^2 = e^2 + (s_0 + s)^2 \qquad (3.1)$$

$$r_B = \sqrt{r_B^2} \qquad (3.2)$$

$$\cos\alpha_B \equiv \sin\tau = \frac{e}{r_B} \qquad (3.3)$$

$$\sin\alpha_B \equiv \cos\tau = \frac{s_0 + s}{r_B} \qquad (3.4)$$

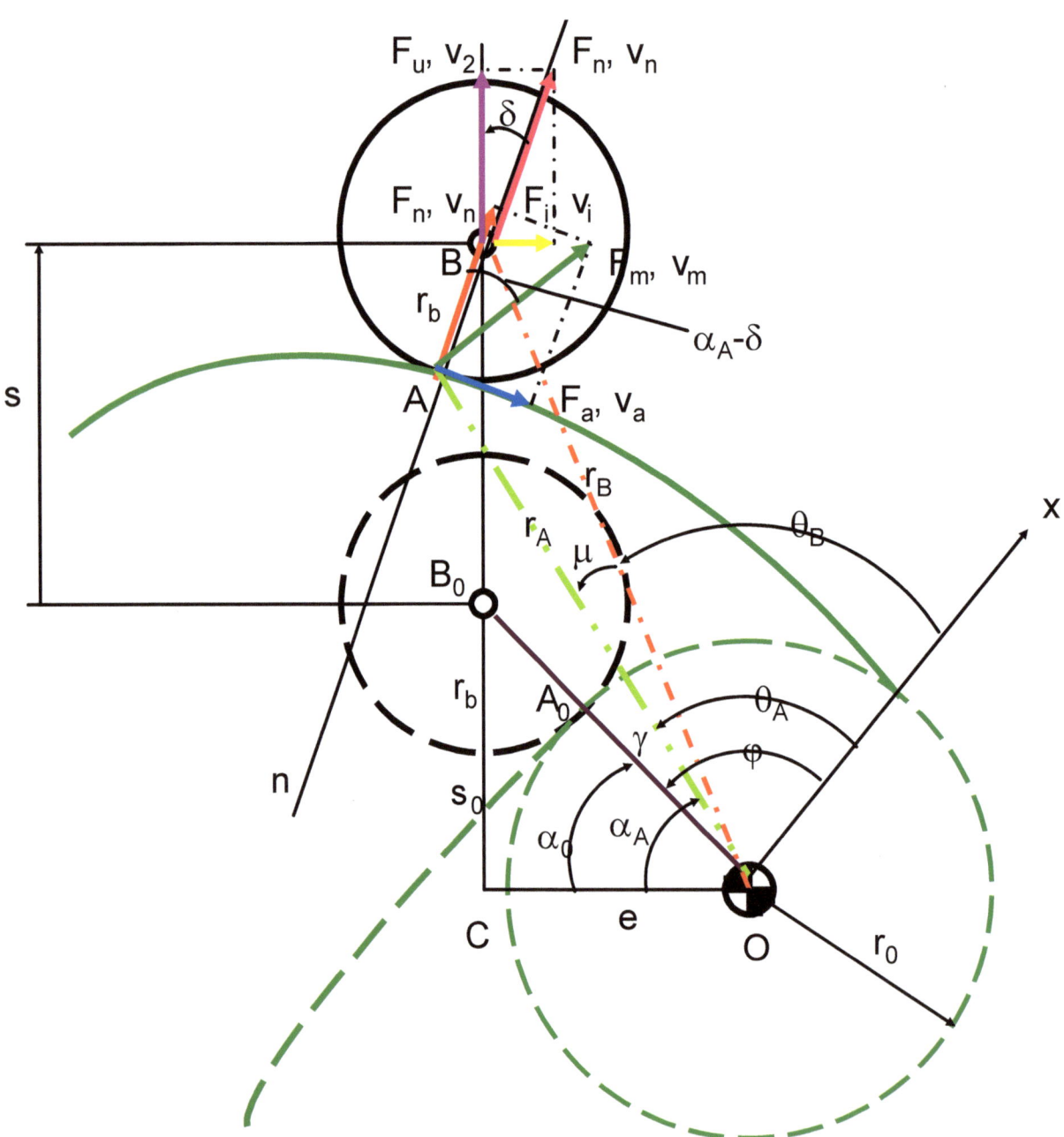

Fig. 2 *Forces and speeds to the cam with translated follower with roll*

$$\cos\delta = \frac{s_0 + s}{\sqrt{(s_0 + s)^2 + (s' - e)^2}} \qquad (3.5)$$

$$\sin\delta = \frac{s'-e}{\sqrt{(s_0+s)^2+(s'-e)^2}} \tag{3.6}$$

$$\cos(\delta+\tau) = \cos\delta \cdot \cos\tau - \sin\delta \cdot \sin\tau \tag{3.7}$$

$$r_A^2 = r_B^2 + r_b^2 - 2 \cdot r_b \cdot r_B \cdot \cos(\delta+\tau) \tag{3.8}$$

$$\cos\alpha_A = \frac{e \cdot \sqrt{(s_0+s)^2+(s'-e)^2} + r_b \cdot (s'-e)}{r_A \cdot \sqrt{(s_0+s)^2+(s'-e)^2}} \tag{3.9}$$

$$\sin\alpha_A = \frac{(s_0+s) \cdot [\sqrt{(s_0+s)^2+(s'-e)^2} - r_b]}{r_A \cdot \sqrt{(s_0+s)^2+(s'-e)^2}} \tag{3.10}$$

$$\cos(\alpha_A-\delta) = \frac{(s_0+s) \cdot s'}{r_A \cdot \sqrt{(s_0+s)^2+(s'-e)^2}} = \frac{s'}{r_A} \cdot \cos\delta \tag{3.11}$$

$$\cos(\alpha_A-\delta) \cdot \cos\delta = \frac{s'}{r_A} \cdot \cos^2\delta \tag{3.12}$$

$$\begin{cases} v_a = v_m \cdot \sin(\alpha_A-\delta) \\ F_a = F_m \cdot \sin(\alpha_A-\delta) \end{cases} \tag{3.13}$$

$$\begin{cases} v_n = v_m \cdot \cos(\alpha_A-\delta) \\ F_n = F_m \cdot \cos(\alpha_A-\delta) \end{cases} \tag{3.14}$$

$$\begin{cases} v_i = v_n \cdot \sin\delta \\ F_i = F_n \cdot \sin\delta \end{cases} \tag{3.15}$$

$$\begin{cases} v_2 = v_n \cdot \cos\delta = v_m \cdot \cos(\alpha_A-\delta) \cdot \cos\delta \\ F_u = F_n \cdot \cos\delta = F_m \cdot \cos(\alpha_A-\delta) \cdot \cos\delta \end{cases} \tag{3.16}$$

$$\begin{cases} P_u = F_u \cdot v_2 = F_m \cdot v_m \cdot \cos^2(\alpha_A-\delta) \cdot \cos^2\delta \\ P_c = F_m \cdot v_m \end{cases} \tag{3.17}$$

$$\eta_i - \frac{P_u}{P_c} - \frac{F_m \cdot v_m \cdot \cos^2(\alpha_A-\delta) \cdot \cos^2\delta}{F_m \cdot v_m} =$$
$$= [\cos(\alpha_A-\delta) \cdot \cos\delta]^2 = [\frac{s'}{r_A} \cdot \cos^2\delta]^2 = \frac{s'^2}{r_A^2} \cdot \cos^4\delta \tag{3.18}$$

4 Determining of momentary dynamic efficiency of the rotary cam and rocking follower with roll

F_m, v_m, are perpendicular to the vector r_A at A. F_m is divided into F_a (the sliding force) and F_n (the normal force). F_n is divided too into F_c (the compressed force) and F_u (the useful force). The written relations are the following [1, 2] (4.1-4.31).

$$\cos\psi_0 = \frac{b^2 + d^2 - (r_0 + r_b)^2}{2 \cdot b \cdot d} \tag{4.1}$$

$$\psi_2 = \psi + \psi_0 \tag{4.2}$$

$$RAD = \sqrt{d^2 + b^2(1-\psi')^2 - 2bd(1-\psi')\cos\psi_2} \tag{4.3}$$

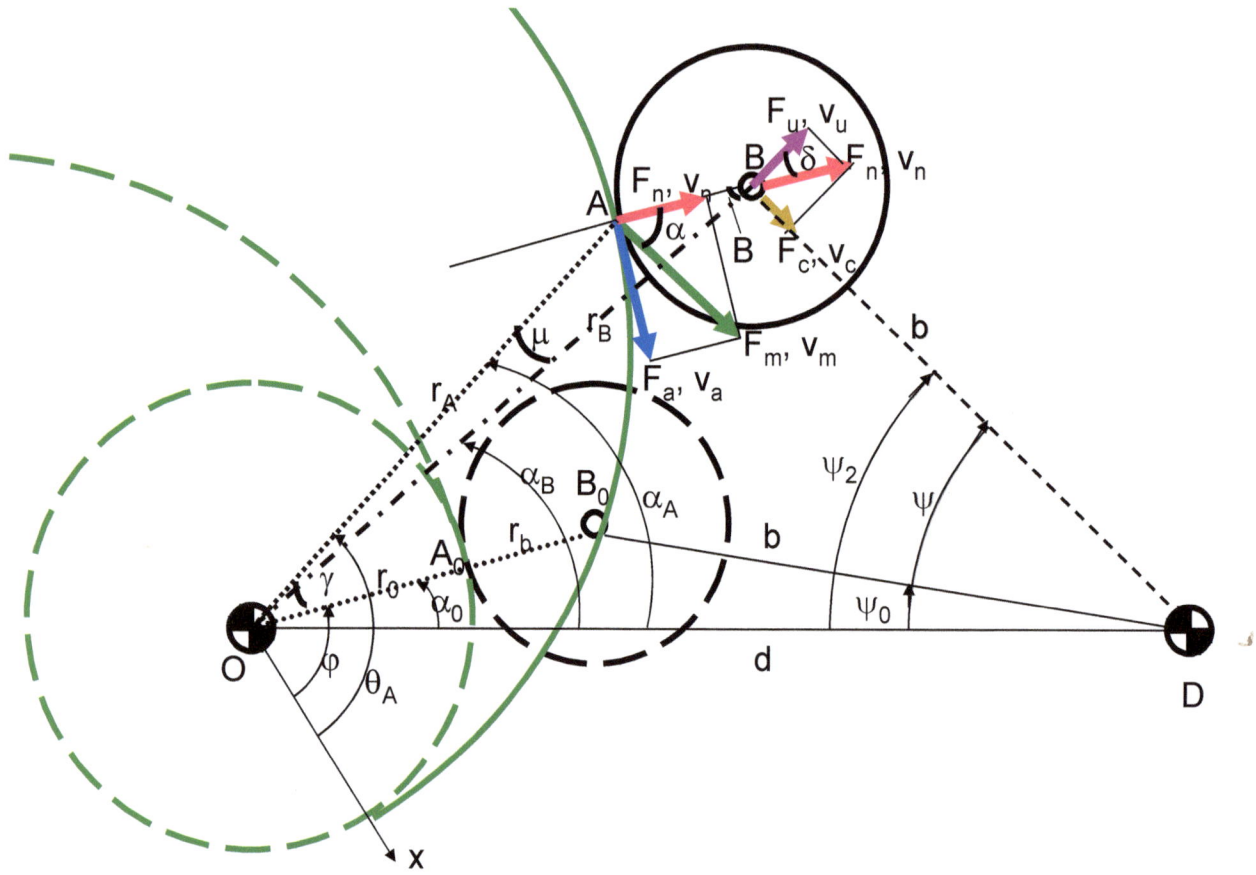

Fig. 3 *Forces and speeds for the rotary cam and rocking follower with roll*

$$\sin\delta = \frac{d \cdot \cos\psi_2 + b \cdot \psi' - b}{RAD} \tag{4.4}$$

$$\cos\delta = \frac{d \cdot \sin\psi_2}{RAD} \tag{4.5}$$

$$r_B^2 = b^2 + d^2 - 2 \cdot b \cdot d \cdot \cos\psi_2 \tag{4.6}$$

$$\cos\alpha_B = \frac{d^2 + r_B^2 - b^2}{2 \cdot d \cdot r_B} \tag{4.7}$$

$$\sin\alpha_B = \frac{b \cdot \sin\psi_2}{r_B} \tag{4.8}$$

$$\sin(\delta + \psi_2) = \sin\delta\cos\psi_2 + \sin\psi_2\cos\delta \tag{4.9}$$

$$\cos(\delta + \psi_2) = \cos\delta\cos\psi_2 - \sin\psi_2\sin\delta \tag{4.10}$$

$$B = \delta + \psi_2 + \alpha_B - \frac{\pi}{2} \tag{4.11}$$

$$\cos B = \sin(\delta + \psi_2 + \alpha_B) \tag{4.12}$$

$$\sin B = -\cos(\delta + \psi_2 + \alpha_B) \tag{4.13}$$

$$\cos B = \sin(\delta + \psi_2) \cdot \cos\alpha_B + \sin\alpha_B \cdot \cos(\delta + \psi_2) \tag{4.14}$$

$$\sin B = \sin(\delta + \psi_2) \cdot \sin\alpha_B - \cos\alpha_B \cdot \cos(\delta + \psi_2) \tag{4.15}$$

$$r_A^2 = r_B^2 + r_b^2 - 2 \cdot r_b \cdot r_B \cdot \cos B \tag{4.16}$$

$$\cos\mu = \frac{r_A^2 + r_B^2 - r_b^2}{2 \cdot r_A \cdot r_B} \tag{4.17}$$

$$\sin\mu = \frac{r_b}{r_A} \cdot \sin B \tag{4.18}$$

$$\alpha_A = \alpha_B + \mu \tag{4.19}$$

$$\cos\alpha_A = \cos\alpha_B\cos\mu - \sin\alpha_B\sin\mu \tag{4.20}$$

$$\sin\alpha_A = \sin\alpha_B\cos\mu + \cos\alpha_B\sin\mu \tag{4.21}$$

$$\alpha = \pi - \alpha_A - \psi_2 - \delta \tag{4.22}$$

$$\cos\alpha = -\cos(\psi_2 + \delta + \alpha_A) =$$
$$= \sin(\psi_2 + \delta) \cdot \sin\alpha_A - \cos(\psi_2 + \delta) \cdot \cos\alpha_A \qquad (4.23)$$

$$\cos\alpha = \frac{\psi' \cdot b}{r_A} \cdot \cos\delta \quad (4.24) \qquad\qquad \cos\alpha \cdot \cos\delta = \frac{\psi' \cdot b}{r_A} \cdot \cos^2\delta \qquad (4.25)$$

$$\begin{cases} F_a = F_m \cdot \sin\alpha \\ v_a = v_m \cdot \sin\alpha \end{cases} \quad (4.26) \qquad\qquad \begin{cases} F_n = F_m \cdot \cos\alpha \\ v_n = v_m \cdot \cos\alpha \end{cases} \qquad (4.27)$$

$$\begin{cases} F_c = F_n \cdot \sin\delta \\ v_c = v_n \cdot \sin\delta \end{cases} \quad (4.28) \qquad \begin{cases} F_u = F_n \cdot \cos\delta = F_m \cdot \cos\alpha \cdot \cos\delta \\ v_2 = v_n \cdot \cos\delta = v_m \cdot \cos\alpha \cdot \cos\delta \end{cases} \qquad (4.29)$$

$$\begin{cases} P_u = F_u \cdot v_2 = F_m \cdot v_m \cdot \cos^2\alpha \cdot \cos^2\delta \\ P_c = F_m \cdot v_m \end{cases} \quad (4.30)$$

$$\eta_i = \frac{P_u}{P_c} = \cos^2\alpha \cdot \cos^2\delta = (\cos\alpha \cdot \cos\delta)^2 = \qquad (4.31)$$
$$= (\frac{\psi' \cdot b}{r_A} \cdot \cos^2\delta)^2 = \frac{\psi'^2 \cdot b^2}{r_A^2} \cdot \cos^4\delta$$

5 Determining of momentary mechanical efficiency of the rotary cam and general plate rocking follower

The written relations are following, (5.1-5.6) (see Fig. 4) [1, 2]:

$$AH = [\sqrt{d^2 - (r_0 - b)^2} \cdot \cos\psi - (r_0 - b) \cdot \sin\psi] \cdot \frac{\psi'}{1 - \psi'} \qquad (5.1)$$

$$OH = b + (r_0 - b) \cdot \cos\psi + \sqrt{d^2 - (r_0 - b)^2} \cdot \sin\psi \qquad (5.2)$$

$$r^2 = AH^2 + OH^2 \qquad (5.3)$$

$$\sin\tau = \frac{AH}{r}; \quad \sin^2\tau = \frac{AH^2}{r^2} = \frac{AH^2}{AH^2 + OH^2} \qquad (5.4)$$

$$\begin{cases} F_n = F_m \cdot \cos\alpha = F_m \cdot \sin\tau; \\ \\ v_n = v_m \cdot \cos\alpha = v_m \cdot \sin\tau \end{cases} \qquad (5.5)$$

$$\eta_i = \frac{P_n}{P_c} = \frac{F_n \cdot v_n}{F_m \cdot v_m} = \frac{F_m \cdot v_m \cdot \sin^2\tau}{F_m \cdot v_m} = \sin^2\tau = \frac{AH^2}{AH^2 + OH^2} \qquad (5.6)$$

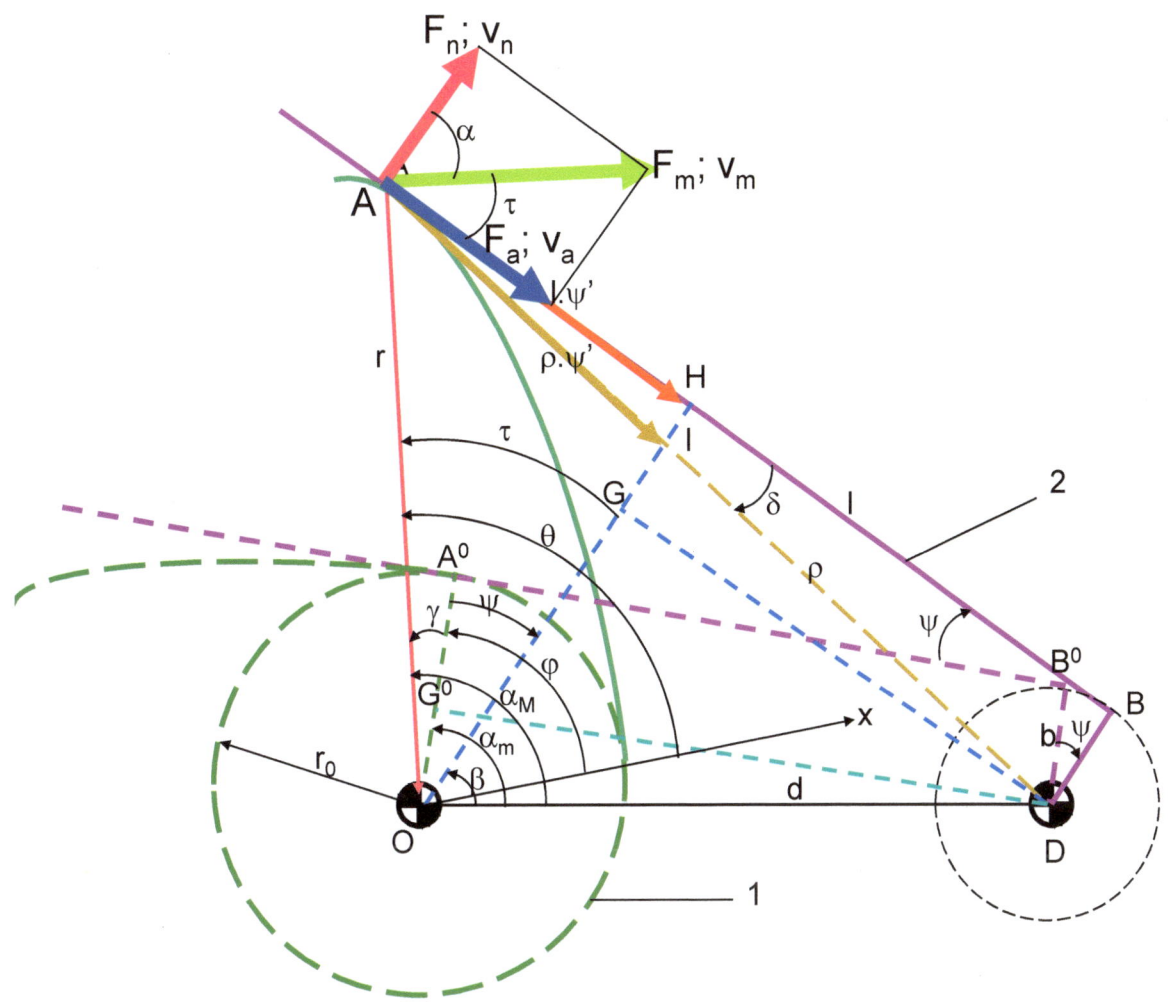

Fig. 4 *Forces and speeds for the rotary cam and general plate rocking follower*

6 Conclusions

The follower with roll makes the input-force be divided into several components. This is the reason why, the dynamics and the precise-kinematics (the dynamic-kinematics) of mechanism with rotary cam and follower with roll, are more different and difficult. The presented dynamic efficiency of followers with roll is not the same like the classical mechanical efficiency. For plate followers the dynamic and the mechanical efficiency are the same. This is the great advantage of plate followers.

References

[1] PETRESCU F.I., PETRESCU R.V., Determining the dynamic efficiency of cams. SYROM 2005, Bucharest, Romania, Vol. I, pp. 129-134, 2005.
[2] PETRESCU F.I., PETRESCU R.V., POPESCU N., The efficiency of cams. In the Second International Conference "Mechanics and Machine Elements", Technical University of Sofia, November 4-6, Sofia, Bulgaria, Vol. II, pp. 237-243, 2005.

CHAPTER II

CONTRIBUTIONS AT THE
DYNAMIC OF CAMS

ABSTRACT: *The chapter presents an original method in determining a general, dynamic and differential equation for the motion of machines and mechanisms, particularized for the mechanisms with rotation cams and followers. This equation can be directly integrated by an original method presented in this chapter. After integration the resulted mother equation may be solved immediately. It presents an original dynamic model with one degree of freedom, with variable internal amortization. It determines the resistant force reduced at the valve (4), the motor force reduced at the valve (5), and the coefficient of variable internal amortization (6). The reduced mass can be calculated with the form (8). The differential motion equation takes the exact form (31), and the approximate form (32). The equation (31) is preparing for its integration with the form (35, 36, 37). The (37) form can be directly integrated and it obtains the parental equation (38). The equation (38) can be arranged in forms (39, 40, 41). The mother equation (41) can be solved directly (42-45), or more elegant with finished differences (48 and 49-50).*

Keywords: Motor-force, resistant-force, variable internal amortization, differential equation, valve rocker, valve push rod, valve lifter, valve spring.

1. INTRODUCTION

The chapter presents shortly an original method in determining a general dynamic differential equation, particularized for the mechanisms with rotation cams and followers [1, 2, 3].

This equation can be integrated directly by an original method presented in this chapter.

2. PRESENTING A DYNAMIC MODEL, WITH ONE GRADE
OF FREEDOM, WITH VARIABLE INTERNAL AMORTIZATION

2.1. Determining the amortization coefficient of the mechanism

Starting with the kinematical schema of the classical valve gear mechanism (see the figure 1), one creates the translating dynamic model, with a single degree of freedom (with a single mass), with variable internal amortization (see the picture 2), having the motion equation (1).

The formula (1) is just a Newton equation, where the sum of forces on a single element is 0, [1, 2, 3]:

$$M \cdot \ddot{x} = K \cdot (y - x) - k \cdot x - c \cdot \dot{x} - F_0 \tag{1}$$

Where:

    M –the mass of the mechanism, reduced at the valve;

    K –the elastically constant of the system;

    k –the elastically constant of the valve spring;

    c –the coefficient of the system's amortization;

    F_0 –the elastically force which compressing the valve spring;

    x –the effective displacement of the valve;

    y≡s –the theoretical displacement of the tappet reduced at the valve, imposed by the cam's profile.

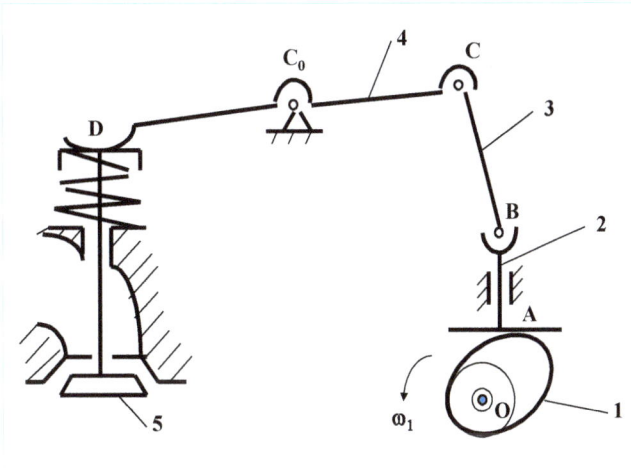

FIG. 1. *The kinematical schema of the classical valve gear mechanism*

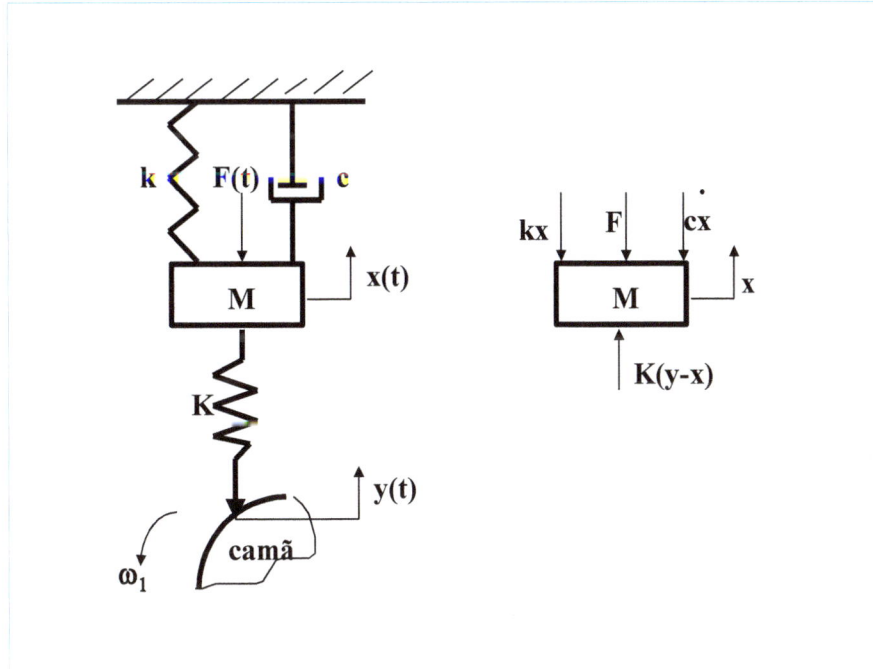

Fig. 2. *Dynamic model with a single liberty, with variable internal amortization*

The Newton equation (1) can be written in form (2):

$$M \cdot \ddot{x} + c \cdot \dot{x} = K \cdot (y - x) - (F_0 + k \cdot x) \tag{2}$$

The differential equation, Lagrange, can be written in form (3).

$$M \cdot \ddot{x} + \frac{1}{2} \cdot \frac{dM}{dt} \cdot \dot{x} = F_m - F_r \tag{3}$$

Comparing the two equations, (2 and 3), we identifie the coefficients and obtain the resistant force (4), the motor force (5) and the coefficient of internal amortization (6), [1, 3]. It can see that the internal amortization coefficient, c, is a variable:

$$F_r = F_0 + k \cdot x = k \cdot x_0 + k \cdot x = k \cdot (x_0 + x) \tag{4}$$

$$F_m = K \cdot (y - x) = K \cdot (s - x) \tag{5}$$

$$c = \frac{1}{2} \cdot \frac{dM}{dt} \tag{6}$$

It places the variable coefficient, c, (see the relation 6), in the Newton equation (form 1 or 2) and obtains the equation (7), [1, 3]:

$$M \cdot \ddot{x} + \frac{1}{2} \cdot \frac{dM}{dt} \cdot \dot{x} + (K + k) \cdot x = K \cdot y - F_0 \tag{7}$$

The reduced mass can be written in form (8), (the reduced mass of the system, reduced at the valve), [1]:

$$M = m_5 + (m_2 + m_3) \cdot (\frac{\dot{y}_2}{\dot{x}})^2 + J_1 \cdot (\frac{\omega_1}{\dot{x}})^2 + J_4 \cdot (\frac{\omega_4}{\dot{x}})^2 \tag{8}$$

With the following notations:

m_2 = the mass of the tappet (of the valve lifter);

m_3 = the mass of the valve push rod;

m_5 = the valve mass;

J_1 = the inertia mechanical moment of the cam;

J_4 = the inertia mechanical moment of the valve rocker;

$\dot{y}_2$ = the tappet velocity, or the second movement-low, imposed by the cam's profile;

$\dot{x}$ = the real (dynamic) valve velocity.

If one notes with $i = i_{25}$, the ratio of transmission tappet-valve, given from the valve rocker, the theoretically velocity of the valve, $\dot{y}$, (the tappet velocity reduced at the valve), takes the form (9), where the ratio of transmission, i, is given from the formula (10).

$$\dot{y} \equiv \dot{y}_5 = \frac{\dot{y}_2}{i} \tag{9}$$

$$i = \frac{CC_0}{C_0 D} \tag{10}$$

It can write the following relations (11-16), where y' is the reduced velocity forced at the tappet by the cam's profile. With the relations (10, 13, 14, 16) the reduced mass (8), can be written in the forms (17–19):

$$\dot{x} = \omega_1 \cdot x' \tag{11}$$

$$\ddot{x} = \omega_1^2 \cdot x'' \tag{12}$$

$$\dot{y}_2 = \omega_1 \cdot \dot{y'}_2 = \omega_1 \cdot i \cdot y' \tag{13}$$

$$\frac{\omega_1}{\dot{x}} = \frac{\omega_1}{\omega_1 \cdot x'} = \frac{1}{x'} \tag{14}$$

$$\omega_4 = \frac{\dot{y}_2}{CC_0} = \frac{\omega_1 \cdot \dot{y'}_2}{CC_0} = \frac{\omega_1 \cdot y' i}{CC_0} = \frac{\omega_1 \cdot y'}{CC_0} \cdot \frac{CC_0}{C_0 D} = \frac{\omega_1 \cdot y'}{C_0 D} \tag{15}$$

$$\frac{\omega_4}{\dot{x}} = \frac{\omega_1 \cdot y'}{C_0 D \cdot \omega_1 \cdot x'} = \frac{1}{C_0 D} \cdot \frac{y'}{x'} \tag{16}$$

$$M = m_5 + (m_2 + m_3) \cdot (\frac{i \cdot y'}{x'})^2 + J_1 \cdot (\frac{1}{x'})^2 + J_4 \cdot (\frac{1}{C_0 D} \cdot \frac{y'}{x'})^2 \tag{17}$$

$$M = m_5 + [i^2 \cdot (m_2 + m_3) + \frac{J_4}{(C_0 D)^2}] \cdot (\frac{y'}{x'})^2 + J_1 \cdot (\frac{1}{x'})^2 \tag{18}$$

$$M = m_5 + m * \cdot (\frac{y'}{x'})^2 + J_1 \cdot (\frac{1}{x'})^2 \tag{19}$$

It derivates dM/dφ and obtains the relations (20–22):

$$\frac{d[(\frac{y'}{x'})^2]}{d\varphi} = \frac{2 \cdot y'}{x'} \cdot \frac{(y'' \cdot x' - x'' \cdot y')}{x'^2} = \tag{20}$$

$$= \frac{2 \cdot y'}{x'^2} \cdot (y'' - x'' \cdot \frac{y'}{x'}) = 2 \cdot (\frac{y'}{x'})^2 \cdot (\frac{y''}{y'} - \frac{x''}{x'})$$

$$\frac{d[(\frac{1}{x'})^2]}{d\varphi} = \frac{2}{x'} \cdot \frac{-x''}{x'^2} = -2 \cdot \frac{x''}{x'^3} \tag{21}$$

$$\frac{dM}{d\varphi} = 2 \cdot m * \cdot (\frac{y'}{x'})^2 \cdot (\frac{y''}{y'} - \frac{x''}{x'}) - 2 \cdot J_1 \cdot \frac{x''}{x'^3} \tag{22}$$

The relation (6) can be written in form (23) and with relation (22), it's taking the forms (24–25):

$$c = \frac{\omega}{2} \cdot \frac{dM}{d\varphi} \tag{23}$$

$$c = \omega \cdot \{[i^2 \cdot (m_2 + m_3) + \frac{J_4}{(C_0 D)^2}] \cdot (\frac{y'}{x'})^2 \cdot (\frac{y''}{y'} - \frac{x''}{x'}) - J_1 \cdot \frac{x''}{x'^3}\} \tag{24}$$

$$c = \omega \cdot [m * \cdot (\frac{y'}{x'})^2 \cdot (\frac{y''}{y'} - \frac{x''}{x'}) - J_1 \cdot \frac{x''}{x'^3}] \tag{25}$$

With the notation (26):

$$m* = i^2 \cdot (m_2 + m_3) + \frac{J_4}{(C_0 D)^2} \tag{26}$$

2.2. Determining the movement equations

With the relations (19, 12, 25, 11) the equation (2) takes the forms (27, 28, 29, 30 and 31):

$$M \cdot \omega^2 \cdot x'' + c \cdot \omega \cdot x' + (K+k) \cdot x = K \cdot y - F_0 \tag{27}$$

$$\omega^2 \cdot x'' \cdot m_5 + \omega^2 \cdot m * \cdot (\frac{y'}{x'})^2 \cdot x'' + J_1 \cdot (\frac{1}{x'})^2 \cdot x'' \omega^2 + \omega^2 \cdot x' \cdot m * \cdot$$
$$(\frac{y'}{x'})^2 \cdot (\frac{y''}{y'} - \frac{x''}{x'}) - x' \cdot \omega^2 \cdot J_1 \cdot \frac{x''}{x'^3} + (K+k) \cdot x = K \cdot y - F_0 \tag{28}$$

$$\omega^2 \cdot m_5 \cdot x'' + \omega^2 \cdot m * \cdot x'' \cdot (\frac{y'}{x'})^2 - \omega^2 \cdot m * \cdot (\frac{y'}{x'})^2 \cdot x'' +$$
$$+ \omega^2 \cdot m * \cdot y'' \cdot \frac{y'}{x'} + (K+k) \cdot x = K \cdot y - F_0 \tag{29}$$

$$\omega^2 \cdot m_5 \cdot x'' + (K+k) \cdot x + \omega^2 \cdot m * \cdot y'' \cdot \frac{y'}{x'} = K \cdot y - F_0 \tag{30}$$

$$\omega^2 \cdot (m_5 \cdot x'' + m * \cdot y'' \cdot \frac{y'}{x'}) + (K+k) \cdot x = K \cdot y - F_0 \tag{31}$$

The exact equation (31) can be approximated at the form (32) with x'≅y':

$$\omega^2 \cdot (m_5 \cdot x'' + m * \cdot y'') + (K+k) \cdot x = K \cdot y - F_0 \tag{32}$$

With the following notations: y=s, y'=s', y''=s'', y'''=s''', the equation (32) takes the approximate form (33) and the complete equation (31) takes the exact form (34).

$$\omega^2 \cdot (m_5 \cdot x'' + m * \cdot s'') + (K+k) \cdot x = K \cdot s - F_0 \tag{33}$$

$$\omega^2 \cdot (m_5 \cdot x'' + m * \cdot s'' \cdot \frac{s'}{x'}) + (K+k) \cdot x = K \cdot s - F_0 \tag{34}$$

3. SOLVING THE DIFFERENTIAL EQUATION BY DIRECT INTEGRATION AND OBTAINING THE MOTHER EQUATION

It integrates the equation (31) directly. It prepares the equation (31) for the integration. First, we write (31) in form (35):

$$-(K+k) \cdot x + K \cdot y - k \cdot x_0 - m_S^* \cdot \omega^2 \cdot x^{II} = \frac{m_T^* \cdot \omega^2 \cdot y^{II} \cdot y^{I}}{x^{I}} \tag{35}$$

The equation (35), can be amplified by x' and obtains the relation (36):

$$-(K+k) \cdot x \cdot x^{I} + K \cdot y \cdot x^{I} - k \cdot x_0 \cdot x^{I} -$$
$$- m_S^* \cdot \omega^2 \cdot x^{I} \cdot x^{II} = m_T^* \cdot \omega^2 \cdot y^{I} \cdot y^{II} \tag{36}$$

Now, it replaces the term K.y.x' with $K \cdot y \cdot \frac{K}{K+k} \cdot y^{I}$, (taken in calculation the statically assumption, $F_m = F_r$) and it obtains the form (37):

$$-(K+k) \cdot x \cdot x^I + \frac{K^2}{K+k} \cdot y \cdot y^I - k \cdot x_0 \cdot x^I - m_S^* \cdot \omega^2 \cdot x^I \cdot x^{II} = m_T^* \cdot \omega^2 \cdot y^I \cdot y^{II} \qquad (37)$$

It integrates directly the equation (37) and obtains the mother equation (38):

$$-(K+k) \cdot \frac{x^2}{2} + \frac{K^2}{K+k} \cdot \frac{y^2}{2} - k \cdot x_0 \cdot x -$$
$$-m_S^* \cdot \omega^2 \cdot \frac{x'^2}{2} = m_T^* \cdot \omega^2 \cdot \frac{y'^2}{2} + C \qquad (38)$$

With the initial condition, at the $\varphi=0$, $y=y'=0$ and $x=x'=0$, it obtains for the constant of integration, C the value 0. In this case the equation (38), takes the form (39):

$$-(K+k) \cdot \frac{x^2}{2} + \frac{K^2}{K+k} \cdot \frac{y^2}{2} - k \cdot x_0 \cdot x - m_S^* \cdot \omega^2 \cdot \frac{x'^2}{2} = m_T^* \cdot \omega^2 \cdot \frac{y'^2}{2} \qquad (39)$$

The equation (39) can be put in the form (40), if one divides it with the $-\dfrac{K+k}{2}$:

$$x^2 + 2 \cdot \frac{k \cdot x_0}{K+k} \cdot x + \frac{m_S^* \cdot \omega^2}{K+k} \cdot x'^2 + \frac{m_T^* \cdot \omega^2}{K+k} y'^2 - \frac{K^2}{(K+k)^2} \cdot y^2 = 0 \qquad (40)$$

The mother equation (40), take the form (41), if one notes: $x' = \dfrac{K}{K+k} \cdot y'$, (the static assumption, $F_m = F_r$).

$$x^2 + 2 \cdot \frac{k \cdot x_0}{K+k} \cdot x - \frac{K^2}{(K+k)^2} \cdot y^2 + \frac{\dfrac{K^2}{(K+k)^2} \cdot m_S^* + m_T^*}{(K+k)} \cdot \omega^2 \cdot y'^2 = 0 \qquad (41)$$

3.1. Solving the mother equation (41) directly

The equation (41) is a two degree equation in x; One determines directly, Δ (42-43) and $X_{1,2}$ (44):

$$\Delta = \frac{(k \cdot x_0)^2 + (K \cdot s)^2}{(K+k)^2} - \frac{m_S^* \cdot \dfrac{K^2}{(K+k)^2} + m_T^*}{(K+k)} \cdot y'^2 \cdot \omega^2 \qquad (42)$$

$$\Delta = \frac{(k \cdot x_0)^2 + (K \cdot s)^2}{(K+k)^2} - \frac{m_S^* \cdot \dfrac{K^2}{(K+k)^2} + m_T^*}{(K+k)} \cdot (D \cdot s')^2 \cdot \omega^2 \qquad (43)$$

$$X_{1,2} = -\frac{k \cdot x_0}{K+k} \pm \sqrt{\Delta} \qquad (44)$$

Physically, just the positive solution is valid (see the relation 45):

$$X = \sqrt{\Delta} - \frac{k \cdot x_0}{K+k} \qquad (45)$$

3.2. Solving the mother equation (41) with finished differences

We can solve the mother equation (41) using the finished differences. We notes:

$$X = s + \Delta X \tag{46}$$

With the notation (46) placed in the mother equation (41), it obtains the equation (47):

$$s^2 + (\Delta X)^2 + 2 \cdot \Delta X \cdot s + 2 \cdot \frac{k \cdot x_0}{K + k} \cdot s + 2 \cdot \frac{k \cdot x_0}{K + k} \cdot \Delta X -$$

$$- \frac{K^2}{(K+k)^2} s^2 + \frac{\dfrac{K^2}{(K+k)^2} \cdot m_S^* + m_T^*}{(K+k)} \cdot \omega^2 \cdot y'^2 = 0 \tag{47}$$

The equation (47) is a two degree equation in ΔX, which can be solved directly with Δ (49) and $\Delta X_{1,2}$, (50), or transformed in a single degree equation in ΔX, with $(\Delta X)^2 \cong 0$, solved by the relation (48).

$$\Delta X = (-1) \cdot \frac{(k^2 + 2 \cdot k \cdot K) \cdot s^2 + 2 \cdot k \cdot x_0 \cdot (K + k) \cdot s + [\dfrac{K^2}{K + k} \cdot m_S^* + (K + k) \cdot m_T^*] \cdot \omega^2 \cdot (Ds')^2}{2 \cdot (s + \dfrac{k \cdot x_0}{K + k}) \cdot (K + k)^2} \tag{48}$$

$$\Delta = \frac{K^2 \cdot s^2 + k^2 \cdot x_0^2 - [\dfrac{K^2}{K + k} \cdot m_S^* + (K + k) \cdot m_T^*] \cdot \omega^2 \cdot (D \cdot s')^2}{(K + k)^2} \tag{49}$$

$$\Delta X = \sqrt{\Delta} - (s + \frac{k \cdot x_0}{K + k}) \tag{50}$$

CONCLUSION

The direct integration of the differential equation (31) generates the mother equation (41), which can be solved directly, with the relation (48). "D" represents the dynamic transmission function (the dynamic transmission coefficient).

REFERENCES

[1] Antonescu, P., Oprean, M., Petrescu, Fl., *Analiza dinamică a mecanismelor de distribuţie cu came*, In: The Proceedings of 7th National Symposium on RIMS, MERO'87, Bucureşti, vol. 3, pp. 126-133, 1987.
[2] Antonescu, P., Petrescu, Fl., *Contributii la analiza cinetoelastodinamică a mecanismelor de distribuţie*, In: The Proceedings of 5th International Symposium on TMM, SYROM'89, Bucureşti, pp. 33-40, 1989.
[3] Petrescu, F., Petrescu, R., *Elemente de dinamica mecanismelor cu came*, In: The Proceedings of 7th National Symposium, PRASIC'02, Braşov, vol. I, pp. 327-332, 2002.

CHAPTER III
CAM GEARS DYNAMICS ILLUSTRATED IN THE CLASSIC DISTRIBUTION

Abstract: *The chapter presents an original method to determine the general dynamics of mechanisms with rotation cams and followers, particularized to the plate translated follower. First, it presents the dynamics kinematics. Then it solves the Lagrange equation and using an original dynamic model with one degree of freedom, with variable internal amortization, it makes the dynamic analysis.*
Keywords: *cam dynamics, classic distribution, cams, followers, dynamics*

1 Introduction

The chapter proposes an original dynamic model illustrated for the rotating cam with plate translate follower. It presents the **dynamics kinematics** (the original kinematics); the variable velocity of the camshaft obtained by an approximate method is used with an original dynamic system having one degree of freedom and a variable internal amortization [1]; it tests two movement laws, one classic and the other original.

2 Dynamics of the classic distribution mechanism
2.1 Precision kinematics in the classic distribution mechanism

In the picture number one, it presents the kinematic schema of the classic distribution mechanism, in two consecutive positions; with an interrupted line is represented the particular position when the follower is situated in the lowest possible plane, (s=0), and the cam which has a clockwise rotation, with constant angular velocity, ω, is situated in the point A^0, (the fillet point between the base profile and the rise profile), a particular point that marks the beginning of the rise movement of the follower, imposed by the cam-profile; with a continue line is represented the higher joint in a certain position of the rise phase.

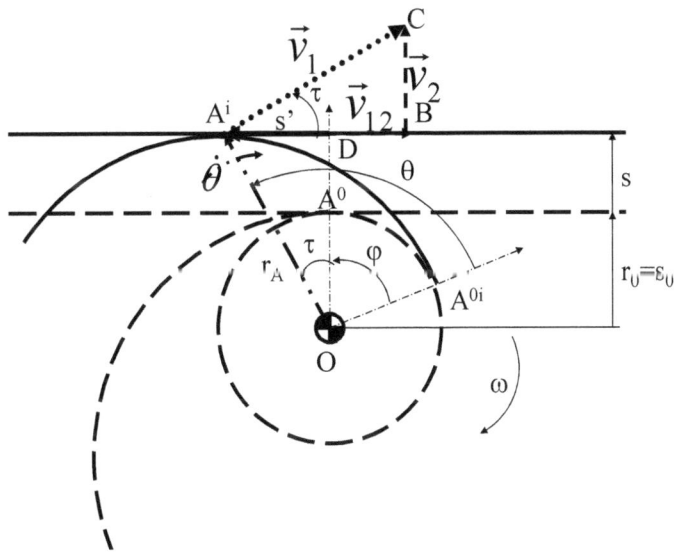

Fig. 1 *The kinematics of the classic distribution mechanism*

The point A^0, which marks the initial higher pair, represents in the same time the contact point between the cam and the follower in the first position. The cam is rotating with the angular velocity, ω (the camshaft angular velocity), describing the angle φ, which shows how the base circle has rotated clockwise (together with the camshaft); this rotation can be seen on the base circle between the two particular points, A^0 and A^{0i}.

In this time the vector $r_A=OA$ (which represents the distance between the centre of cam O, and the contact point A), has rotated anticlockwise with the angle τ. If one measures the angle θ, which positions the general vector, r_A, in function of the particular vector, r_{A0}, it obtains the relation (0):

$$\theta = \varphi + \tau \tag{0}$$

where r_A is the module of the vector $\vec{r}_A$, and θ_A represents the phase angle of the vector $\vec{r}_A$.

The angular velocity of the vector $\vec{r}_A$ is $\dot{\theta}_A$ which is a function of the angular velocity of the camshaft, ω, and of the angle φ (by the movement laws s(φ), s'(φ), s''(φ)).

The follower isn't acted directly by the angle φ and the angular velocity ω; it's acted by the vector $\vec{r}_A$, which has the module r_A, the position angle θ_A and the angular velocity $\dot{\theta}_A$. From here we deduce a particular (dynamic) kinematics, the classical kinematics being just static and approximate kinematics.

Kinematic, it defines the next velocities (Fig. 1).

$\vec{v}_1$ =the cam's velocity; which is the velocity of the vector $\vec{r}_A$, in the point A; now the classical relation (1) becomes an approximate relation, and the real relation takes the form (2).

$$v_1 = r_A.\omega \tag{1}$$

$$v_1 = r_A.\dot{\theta}_A \tag{2}$$

The velocity $\vec{v}_1 = AC$ is separating into the velocity $\vec{v}_2$ =BC (the follower's velocity which acts on its axe, vertically) and $\vec{v}_{12}$ =AB (the slide velocity between the two profiles, the sliding velocity between the cam and the follower, which works along the direction of the commune tangent line of the two profiles in the contact point).

Because usually the cam profile is synthesis for the classical module C with the AD=s' known, we can write the relations:

$$r_A^2 = (r_0 + s)^2 + s'^2 \tag{3}$$

$$r_A = \sqrt{(r_0 + s)^2 + s'^2} \tag{4}$$

$$\cos \tau = \frac{r_0 + s}{r_A} = \frac{r_0 + s}{\sqrt{(r_0 + s)^2 + s'^2}} \tag{5}$$

$$\sin \tau = \frac{AD}{r_A} = \frac{s'}{r_A} = \frac{s'}{\sqrt{(r_0 + s)^2 + s'^2}} \tag{6}$$

$$v_2 = v_1.\sin \tau = r_A.\dot{\theta}_A.\frac{s'}{r_A} = s'.\dot{\theta}_A \tag{7}$$

Now, the follower's velocity isn't $\dot{s}$ ($v_2 \neq \dot{s} \equiv s' \cdot \omega$), but it's given by the relation (9). In the case of the classical distribution mechanism the transmitting function D is given by the relations (8):

$$\dot{\theta}_A = D.\omega$$

$$D = \frac{\dot{\theta}_A}{\omega} \tag{8}$$

$$v_2 = s'.\dot{\theta}_A = s'.D.\omega \tag{9}$$

The determining of the sliding velocity between the profiles is made with the relation (10):

$$v_{12} = v_1.\cos\tau = r_A.\dot{\theta}_A.\frac{r_0 + s}{r_A} = (r_0 + s).\dot{\theta}_A \tag{10}$$

The angles τ and θ_A will be determined, and also their first and second derivatives.

The τ angle has been determined from the triangle ODA^i (Fig.1) with the relations (11-13):

$$\sin\tau = \frac{s'}{\sqrt{(r_0 + s)^2 + s'^2}} \tag{11}$$

$$\cos\tau = \frac{r_0 + s}{\sqrt{(r_0 + s)^2 + s'^2}} \tag{12}$$

$$tg\tau = \frac{s'}{r_0 + s} \tag{13}$$

It derives (11) in function of φ angle and obtains (14):

$$\tau'.\cos\tau = \frac{s''.r_A - s'.\dfrac{(r_0 + s).s' + s'.s''}{r_A}}{(r_0 + s)^2 + s'^2} \tag{14}$$

The relation (14) will be written in the form (15):

$$\tau'.\cos\tau = \frac{s''.(r_0 + s)^2 + s''.s'^2 - s'^2.(r_0 + s) - s'^2.s''}{[(r_0 + s)^2 + s'^2].\sqrt{(r_0 + s)^2 + s'^2}} \tag{15}$$

From the relation (12) it extracts the value of $\cos\tau$, which will be introduced in the left term of the expression (15); then we can reduce $s''.s'^2$ from the right term of the expression (15) and it obtains the relation (16):

$$\tau'.\frac{r_0 + s}{\sqrt{(r_0 + s)^2 + s'^2}} = \frac{(r_0 + s).[s''.(r_0 + s) - s'^2]}{[(r_0 + s)^2 + s'^2].\sqrt{(r_0 + s)^2 + s'^2}} \tag{16}$$

After some simplifications the relation (17), which represents the expression of τ', is finally obtained:

$$\tau' = \frac{s''.(r_0 + s) - s'^2}{(r_0 + s)^2 + s'^2} \tag{17}$$

23

Now when τ' has been explicitly deduced, the next derivatives can be determined. The expression (17) will be derived directly and it obtains for the beginning the relation (18):

$$\tau'' = \frac{[s'''(r_0 + s) + s''s' - 2s's''][(r_0 + s)^2 + s'^2] - 2[s''(r_0 + s) - s'^2][(r_0 + s)s' + s's'']}{[(r_0 + s)^2 + s'^2]^2} \qquad (18)$$

The terms from the first bracket of the numerator (s'.s") are reduced, and then it draws out s' from the fourth bracket of the numerator and obtains the expression (19):

$$\tau'' = \frac{[s'''.(r_0 + s) - s'.s''].[(r_0 + s)^2 + s'^2] - 2.s'.[s''.(r_0 + s) - s'^2].[r_0 + s + s'']}{[(r_0 + s)^2 + s'^2]^2} \qquad (19)$$

Now we can calculate θ_A, with its first two derivatives, $\dot{\theta}_A$ and $\ddot{\theta}_A$. We will write θ instead of θ_A, to simplify the notation. It determines the relation (20) which is the same of (0):

$$\theta = \tau + \varphi \qquad (20)$$

We derive the relation (20) and one obtains the expression (21):

$$\dot{\theta} = \dot{\tau} + \dot{\varphi} = \tau'.\omega + \omega = \omega.(1 + \tau') = D.\omega \qquad (21)$$

It derives twice (20), or derives (21) and obtains (22):

$$\ddot{\theta} = \ddot{\tau} + \ddot{\varphi} = \tau''.\omega^2 = D'.\omega^2 \qquad (22)$$

We can write now the transmission functions, D and D' (for the classical module, C), in the forms (23-24):

$$D = \tau' + 1 \qquad (23)$$

$$D^I = \tau'' \qquad (24)$$

To calculate the follower's velocity (25) we need the expression of the transmission function, D.

$$v_2 = s'\cdot w = s'\cdot\dot{\theta}_A = s'\cdot\dot{\theta} = s'\cdot D\cdot\omega = \dot{s}\cdot D \qquad (25)$$

Where:

$$w = D.\omega \qquad (26)$$

For the classical distribution mechanism (Module C), the variable w is the same as $\dot{\theta}_A$ (see the relation 25).

But in the case of B and F modules (at the cam gears where the follower has a roll), the transmitted function D and w take complex forms.

We can determine now the acceleration of the follower (27).

$$\ddot{y} \equiv a_2 = (s''\cdot D + s'\cdot D')\cdot\omega^2 \qquad (27)$$

Figure 2 represents the classical and dynamic kinematics; the velocities (a), and the accelerations (b).

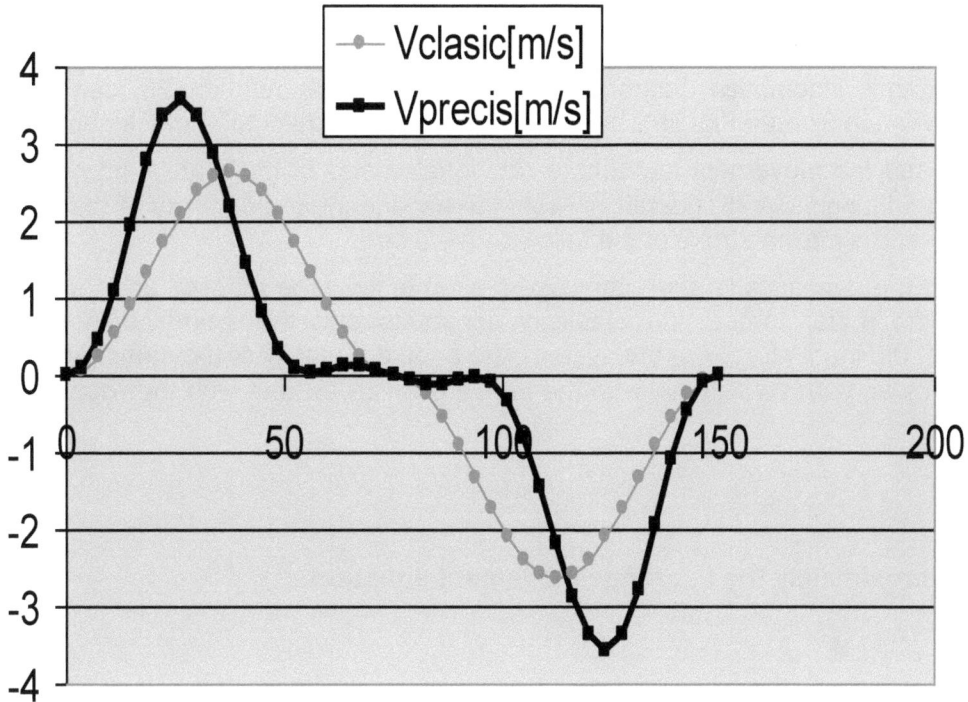

Fig. 2a *The classical and dynamic kinematics; velocities of the follower*

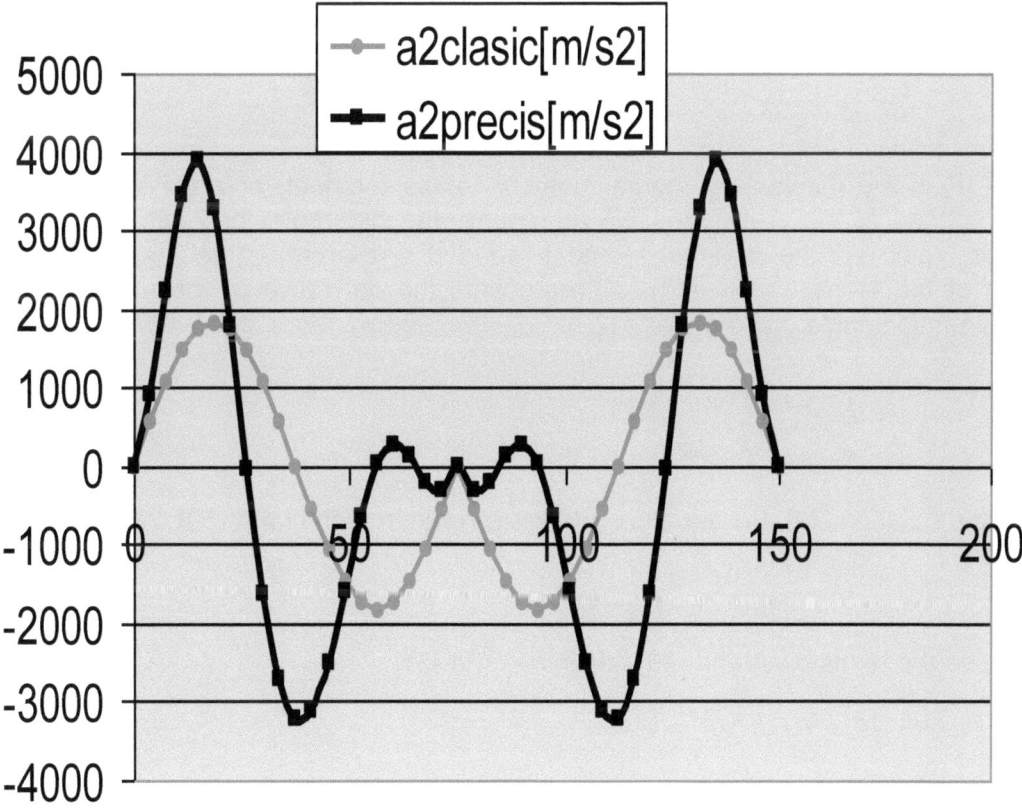

Fig. 2b *The classical and dynamic kinematics; accelerations of the follower*

To determine the acceleration of the follower, s' and s", D and D', τ' and τ" are necessary be known.

The dynamic kinematics diagrams of v_2 (obtained with relation 25, see Fig. 2a), and a_2 (obtained with relation 27, see Fig. 2b), have a more dynamic aspect than one kinematic (classic).

It has used the movement law SIN, a rotational speed of the crankshaft n=5500 rpm, a rise angle φ_u=75⁰, a fall angle φ_d=75⁰ (identically with the ascendant angle), a ray of the basic circle of the cam, r_0=17 mm and a maxim stroke of the follower, h_T=6 mm.

Anyway, the dynamics is more complex, having in view the masses and the inertia moments, the resistant and motor forces, the elasticity constants and the amortization coefficient of the kinematic chain, the inertia forces of the system, the angular velocity of the camshaft and the variation of the camshaft's angular velocity, ω, with the cam's position, φ, and with the rotational speed of the crankshaft, n.

2.2 Solving approximately the Lagrange movement equation

In the kinematics and the static forces study of the mechanisms one considers the shaft's angular velocity constant, $\dot{\varphi} = \omega$ =constant, and the angular acceleration null, $\ddot{\varphi} = \dot{\omega} = \varepsilon = 0$. In reality, this angular velocity ω isn't constant, it is variable with the camshaft position, φ.

In mechanisms with cam and follower the camshaft's angular velocity is variable as well.We shall see further the Lagrange equation, written in the differentiate mode and its general solution.The differentiate Lagrange equation has the form (28):

$$J^* . \ddot{\varphi} + \frac{1}{2} . J^{*I} . \dot{\varphi}^2 = M^* \qquad (28)$$

Where J* is the mechanical inertia moment (mass moment, or mechanic moment) of the mechanism, reduced at the crank, and M* represents the difference between the motor moment reduced at the crank and the resistant moment reduced at the crank; the angle φ represents the rotation angle of the crank (crankshaft). J^{*I} represents the derivative of the mechanic moment in function of the rotation angle φ of the crank (29).

$$\frac{1}{2} . J^{*I} = \frac{1}{2} . \frac{dJ^*}{d\varphi} = L \qquad (29)$$

Using the notation (29), the equation (28) will be written in the form (30):

$$J^* . \ddot{\varphi} + L . \dot{\varphi}^2 = M^* \qquad (30)$$

We divide the terms by J* and (30) takes the form (31):

$$\ddot{\varphi} + \frac{L}{J^*} . \dot{\varphi}^2 = \frac{M^*}{J^*} \qquad (31)$$

The term with $\dot{\varphi}^2$ will be moved to the right side of the equation and the form (32) will be obtained:

$$\ddot{\varphi} = \frac{M^*}{J^*} - \frac{L}{J^*} \cdot \dot{\varphi}^2 \tag{32}$$

Replacing the left term of the expression (32) with (33) we obtain the relation (34):

$$\ddot{\varphi} = \frac{d\dot{\varphi}}{dt} = \frac{d\dot{\varphi}}{d\varphi} \cdot \frac{d\varphi}{dt} = \frac{d\dot{\varphi}}{d\varphi} \cdot \dot{\varphi} = \frac{d\omega}{d\varphi} \cdot \omega \tag{33}$$

$$\omega \cdot \frac{d\omega}{d\varphi} = \frac{M^*}{J^*} - \frac{L}{J^*} \cdot \omega^2 = \frac{M^* - L \cdot \omega^2}{J^*} \tag{34}$$

Because, for an angle φ, ω is different from the nominal constant value ω_n, it can write the relation (35), where $d\omega$ represents the momentary variation for the angle φ; the variable $d\omega$ and the constant ω_n lead us to the needed variable, ω:

$$\omega = \omega_n + d\omega \tag{35}$$

In the relation (35), ω and $d\omega$ are functions of the angle φ, and ω_n is a constant parameter, which can take different values in function of the rotational speed of the drive-shaft, n. At a moment, n is a constant and ω_n is a constant as well (because ω_n is a function of n). The angular velocity, ω, becomes a function of n too (see the relation 36):

$$\omega(\varphi, n) = \omega_n(n) + d\omega(\varphi, \omega_n(n)) \tag{36}$$

With (35) in (34), it obtains the equation (37):

$$(\omega_n + d\omega) \cdot d\omega = [\frac{M^*}{J^*} - \frac{L}{J^*} \cdot (\omega_n + d\omega)^2] \cdot d\varphi \tag{37}$$

The relation (37) takes the form (38):

$$\omega_n \cdot d\omega + (d\omega)^2 = \frac{M^*}{J^*} \cdot d\varphi - \frac{L}{J^*} \cdot d\varphi \cdot [\omega_n^2 + (d\omega)^2 + 2 \cdot \omega_n \cdot d\omega] \tag{38}$$

The equation (38) will be written in the form (39):

$$\omega_n \cdot d\omega + (d\omega)^2 - \frac{M^*}{J^*} \cdot d\varphi + \frac{L}{J^*} \cdot d\varphi \cdot \omega_n^2 +$$
$$+ \frac{L}{J^*} \cdot d\varphi \cdot (d\omega)^2 + 2 \cdot \frac{L}{J^*} \cdot d\varphi \cdot \omega_n \cdot d\omega = 0 \tag{39}$$

The relation (39) takes the form (40):

$$(\frac{L}{J^*} \cdot d\varphi + 1) \cdot (d\omega)^2 + 2 \cdot (\frac{L}{J^*} \cdot d\varphi + \frac{1}{2}) \cdot \omega_n \cdot d\omega -$$
$$(\frac{M^*}{J^*} \cdot d\varphi - \frac{L}{J^*} \cdot d\varphi \cdot \omega_n^2) = 0 \tag{40}$$

The relation (40) is an equation of the second degree in $d\omega$. The discriminate of the equation (40) can be written in the forms (41) and (42):

$$\Delta = \frac{L^2}{J^{*2}} \cdot (d\varphi)^2 \cdot \omega_n^2 + \frac{\omega_n^2}{4} + \frac{L}{J^*} \cdot d\varphi \cdot \omega_n^2 + \frac{L \cdot M^*}{J^{*2}} \cdot (d\varphi)^2$$
$$+ \frac{M^*}{J^*} \cdot d\varphi - \frac{L^2}{J^{*2}} \cdot (d\varphi)^2 \cdot \omega_n^2 - \frac{L}{J^*} \cdot d\varphi \cdot \omega_n^2 \tag{41}$$

27

$$\Delta = \frac{\omega_n^2}{4} + \frac{L.M^*}{J^{*2}}.(d\varphi)^2 + \frac{M^*}{J^*}.d\varphi \tag{42}$$

We keep for dω just the positive solution, which can generate positives and negatives normal values (43), and in this mode only normal values will be obtained for ω; for $\Delta < 0$ it considers dω=0 (this case must be not seeing if the equation is correct).

$$d\omega = \frac{-\dfrac{L}{J^*}.d\varphi.\omega_n - \dfrac{\omega_n}{2} + \sqrt{\Delta}}{\dfrac{L}{J^*}.d\varphi + 1} \tag{43}$$

Observations: For mechanisms with rotate cam and follower, using the new relations, with M* (the reduced moment of the mechanism) obtained by the writing of the known reduced resistant moment and by the determination of the reduced motor moment by the integration of the resistant moment it frequently obtains some bigger values for dω, or zones with Δ negative, with complex solutions for dω. This fact gives us the obligation to reconsider the method to determine the reduced moment.

If we take into consideration M*_r and M*_m, calculated independently (without integration), it obtains for the mechanisms with cam and follower normal values for dω, and $\Delta \geq 0$.

In paper [1] it presents the relations to determine the resistant force (44) reduced to the valve, and the motor force (45) reduced to the ax of the valve:

$$F_r^* = k.(x_0 + x) \tag{44}$$

$$F_m^* = K.(y - x) \tag{45}$$

The reduced resistant moment (46), or the reduced motor moment (47), can be obtained by the resistant or motor force multiplied by the reduced velocity, x'.

$$M_r^* = k.(x_0 + x).x' \tag{46}$$

$$M_m^* = K.(y - x).x' \tag{47}$$

2.3 The dynamic relations used

The dynamics relations used (48-49), have been deduced in the paper [1]:

$$\Delta X = (-1) \cdot \frac{(k^2 + 2 \cdot k \cdot K) \cdot s^2 + 2 \cdot k \cdot x_0 \cdot (K+k) \cdot s + [\dfrac{K^2}{K+k} \cdot m_S^* + (K+k) \cdot m_T^*] \cdot \omega^2 \cdot (Ds')^2}{2 \cdot (s + \dfrac{k \cdot x_0}{K+k}) \cdot (K+k)^2} \tag{48}$$

$$X = s - \frac{[\dfrac{K^2}{K+k} \cdot m_S^* + (K+k) \cdot m_T^*] \cdot \omega^2 \cdot (Ds')^2}{2 \cdot (s + \dfrac{k \cdot x_0}{K+k}) \cdot (K+k)^2}$$
$$- \frac{(k^2 + 2 \cdot k \cdot K) \cdot s^2 + 2 \cdot k \cdot x_0 \cdot (K+k) \cdot s}{2 \cdot (s + \dfrac{k \cdot x_0}{K+k}) \cdot (K+k)^2} \tag{49}$$

28

2.4 The dynamic analysis

The dynamic analysis or the classical movement law sin, can be seen in the diagram from figure 3, and in figure 4 one can see the diagram of an original movement law (C4P) (module C).

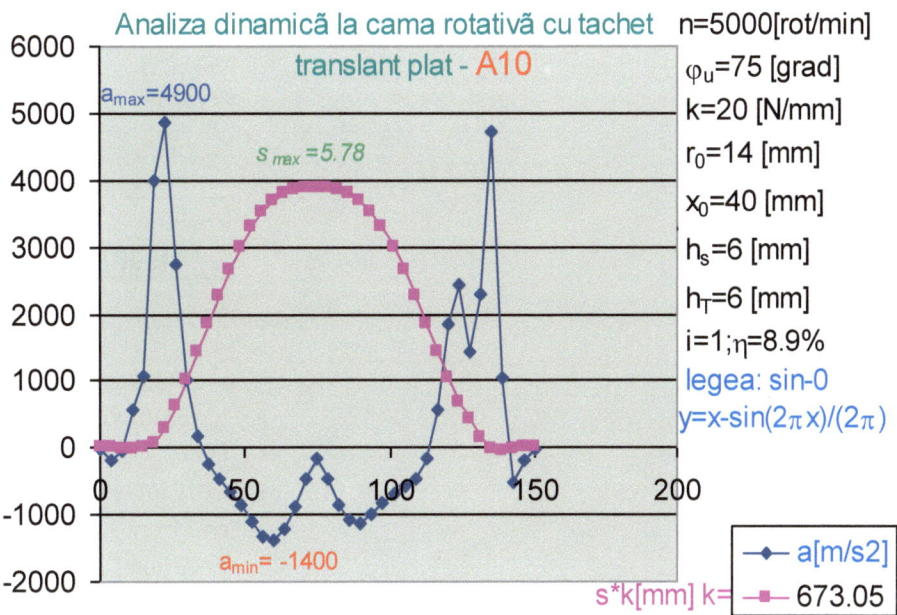

Fig. 3 *The dynamic analysis of the law sin, Module C, $\varphi_u=75^0$, n=5000 rpm*

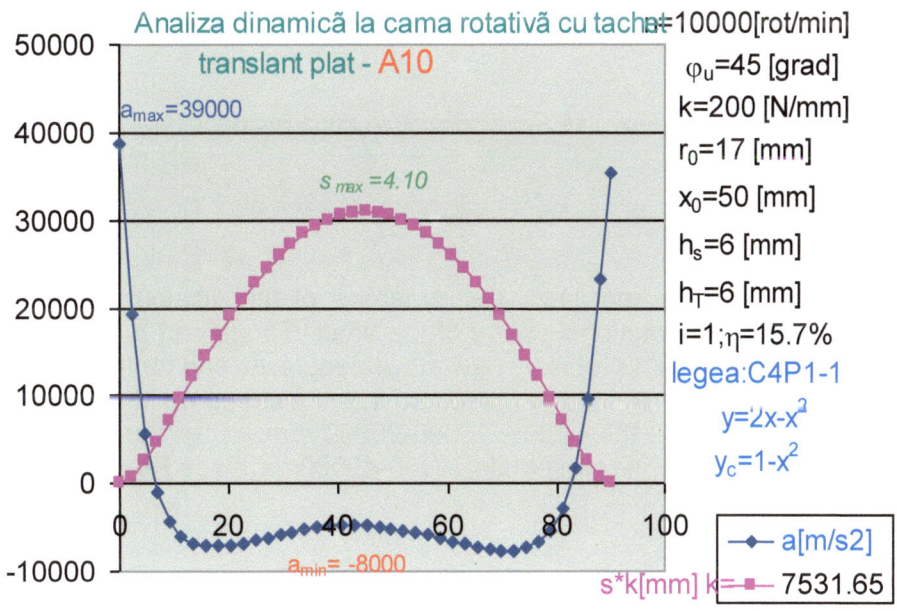

Fig. 4 *The dynamic analysis of the new law, C4P, Module C, $\varphi_u=45^0$, n=10000 rpm*

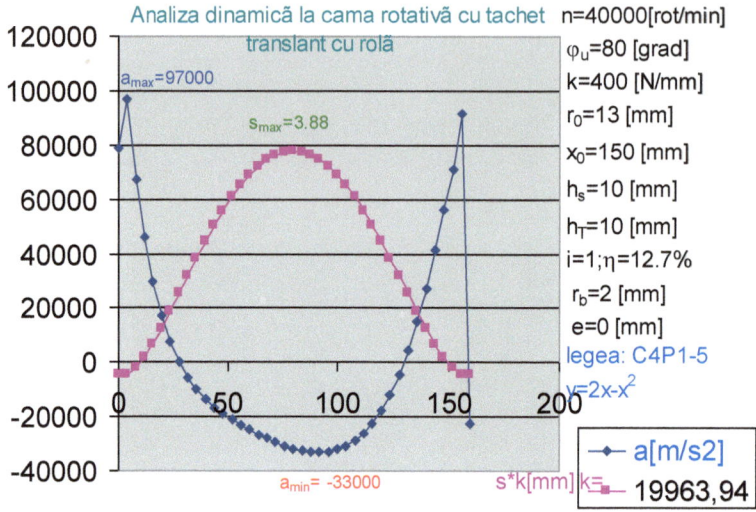

Fig. 5 *Law C4P1-5, Module B, $\varphi_u=80^0$, n=40000 rpm*

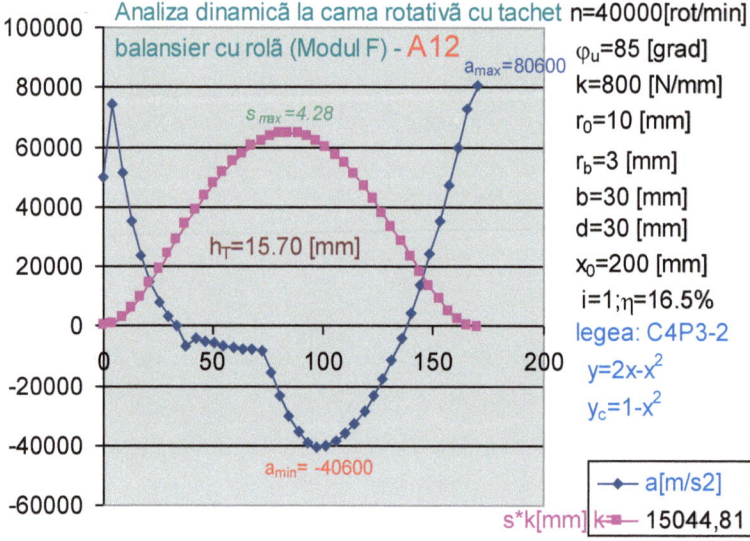

Fig. 6 *Law C4P3-2, Module F, $\varphi_u=85^0$, n=40000 rpm*

3 Conclusions

Using the classical movement laws, the dynamics of the distribution cam-gears depreciate rapidly at the increasing of the rotational speed of the shaft. To support a high rotational speed it is necessary the synthesis of the cam-profile by new movement laws, and for the new Modules.

A new and original movement law is presented in the pictures number 4, 5 and 6; it allows the increase of the rotational speed to the values: 10000-20000 rpm, in the classical module C presented (Fig. 4). With others modules (B, F) it can obtain 30000-40000 rpm (see Figs. 5, 6).

References

[1] Petrescu F.I., Petrescu R.V., *Contributions at the dynamics of cams*. In the Ninth IFToMM International Sympozium on Theory of Machines and Mechanisms, SYROM 2005, Bucharest, Romania, Vol. I, pp. 123-128, 2005.

CHAPTER IV
CAM GEARS DYNAMICS TO THE MODULE B
(WITH TRANSLATED FOLLOWER WITH ROLL)

Abstract: *The chapter briefly presents an original method for determining the dynamics of mechanisms with rotation cam and translated follower with roll. First, one presents the dynamics kinematics. Then one performs the dynamic analysis of a few models, for some movement laws, imposed on the follower, by the designed cam profile.*
Keywords: *cam dynamics, translated follower with roll, movement laws, dynamics kinematic*

1 Introduction

The chapter proposes an original dynamic model of the cam gear with a translated follower with a roll. First, one presents the *dynamics kinematics.* Then one performs the dynamic analysis of a few models, for some movement laws, imposed on the follower, by the designed cam profile.

2 The dynamics of distribution mechanisms with translated follower with roll
2.1 Generalities

The angle α_0 defines the basic position of the vector, $\bar{r}_{B0}$, in the OCB_0 triangle having a right angle (1-4):

$$r_{B_0} = r_0 + r_b \quad (1)$$

$$s_0 = \sqrt{r_{B_0}^2 - e^2} \qquad (2)$$

$$\cos \alpha_0 = \frac{e}{r_{B_0}} \quad (3)$$

$$\sin \alpha_0 = \frac{s_0}{r_{B_0}} \qquad (4)$$

The pressure angle, δ, between the normal n (which passes through the contact point A) and a vertical line, can be calculated with relations (5-7).

$$\cos \delta = \frac{s_0 + s}{\sqrt{(s_0 + s)^2 + (s'-e)^2}} \qquad (5)$$

$$\sin \delta = \frac{s'-e}{\sqrt{(s_0 + s)^2 + (s'-e)^2}} \qquad (6)$$

$$tg\delta = \frac{s'-e}{s_0 + s} \qquad (7)$$

The vector $\bar{r}_A$ can be determined with relations (8-9):

$$r_A^2 = (e + r_b \cdot \sin \delta)^2 + (s_0 + s - r_b \cdot \cos \delta)^2 \qquad (8)$$

$$r_A = \sqrt{(e + r_b \cdot \sin \delta)^2 + (s_0 + s - r_b \cdot \cos \delta)^2} \qquad (9)$$

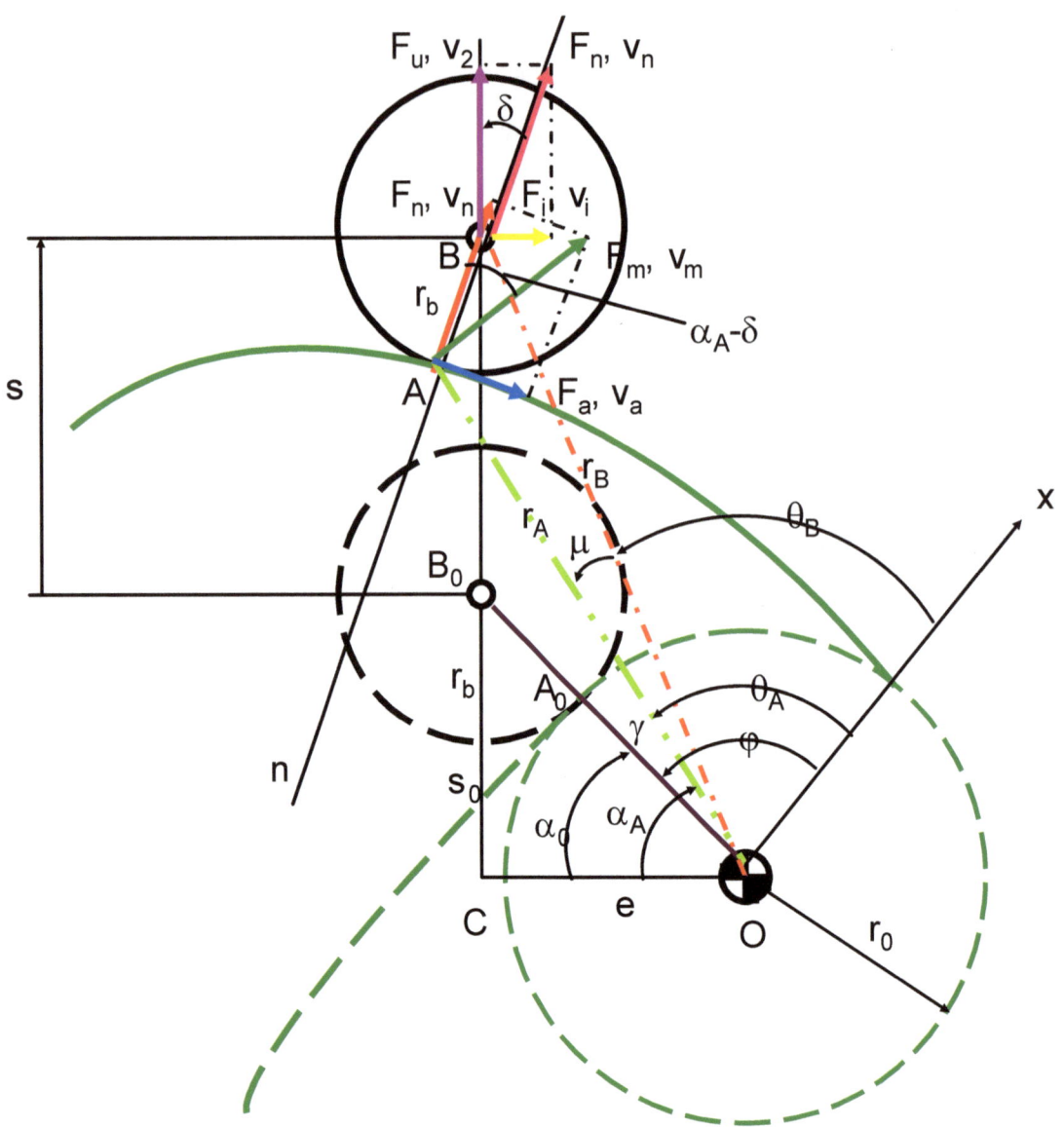

Fig. 1 *Mechanism with rotating cam and translating follower with roll*

We can calculate α_A (10-11):

$$\cos \alpha_A = \frac{e + r_b \cdot \sin \delta}{r_A} \tag{10}$$

$$\sin \alpha_A = \frac{s_0 + s - r_b \cdot \cos \delta}{r_A} \tag{11}$$

2.2 The relations to design the profile

$$\gamma = \alpha_A - \alpha_0 \tag{12}$$

$$\cos \gamma = \cos \alpha_A \cdot \cos \alpha_0 + \sin \alpha_A \cdot \sin \alpha_0 \tag{13}$$

$$\sin \gamma = \sin \alpha_A \cdot \cos \alpha_0 - \cos \alpha_A \cdot \sin \alpha_0 \tag{14}$$

$$\theta_A = \varphi - \gamma \tag{15}$$

$$\cos \theta_A = \cos \varphi \cdot \cos \gamma + \sin \varphi \cdot \sin \gamma \tag{16}$$

$$\sin \theta_A = \sin \varphi \cdot \cos \gamma - \sin \gamma \cdot \cos \varphi \tag{17}$$

2.3 The exact kinematics of B Module

From the triangle OCB (fig. 1) the length r_B (OB) and the complementary angles α_B (COB) and τ (CBO) are determined.

$$r_B^2 = e^2 + (s_0 + s)^2 \tag{18}$$

$$r_B = \sqrt{r_B^2} \tag{19}$$

$$\cos \alpha_B \equiv \sin \tau = \frac{e}{r_B} \tag{20}$$

$$\sin \alpha_B \equiv \cos \tau = \frac{s_0 + s}{r_B} \tag{21}$$

From the general triangle OAB, where one knows OB, AB, and the angle between them, B (ABO, which is the sum of τ and δ), the length OA and the angle μ (AOB) can be determined:

$$\cos(\delta + \tau) = \cos \delta \cdot \cos \tau - \sin \delta \cdot \sin \tau \tag{22}$$

$$r_A^2 = r_B^2 + r_b^2 - 2 \cdot r_b \cdot r_B \cdot \cos(\delta + \tau) \tag{23}$$

$$\cos \mu = \frac{r_A^2 + r_B^2 - r_b^2}{2 \cdot r_A \cdot r_B} \tag{24}$$

$$\sin(\delta + \tau) = \sin \delta \cdot \cos \tau + \sin \tau \cdot \cos \delta \tag{25}$$

$$\sin \mu = \frac{r_b}{r_A} \cdot \sin(\delta + \tau) \tag{26}$$

With α_B and μ we can deduce now α_A and $\dot{\alpha}_A$:

$$\alpha_A = \alpha_B - \mu \tag{27}$$

$$\dot{\alpha}_A = \dot{\alpha}_B - \dot{\mu} \tag{28}$$

From (20) one obtains $\dot{\alpha}_B$ (32), (see 29-32) where $\dot{r}_B$ (31) can be deduced from (18). Then, (33) will be obtained from (24):

$$-\sin \alpha_B \cdot \dot{\alpha}_B = -\frac{e \cdot \dot{r}_B}{r_B^2} \tag{29}$$

$$\dot{\alpha}_B = \frac{e \cdot r_B \cdot \dot{r}_B}{(s_0 + s) \cdot r_B^2} \tag{30}$$

$$2 \cdot r_B \cdot \dot{r}_B = 2 \cdot (s_0 + s) \cdot \dot{s} \qquad r_B \cdot \dot{r}_B = (s_0 + s) \cdot \dot{s} \tag{31}$$

$$\dot{\alpha}_B = \frac{e \cdot (s_0 + s) \cdot \dot{s}}{(s_0 + s) \cdot r_B^2} = \frac{e \cdot \dot{s}}{r_B^2} \tag{32}$$

$$2 \cdot \dot{r}_A \cdot r_B \cdot \cos\mu + 2 \cdot r_A \cdot \dot{r}_B \cdot \cos\mu - $$
$$- 2 \cdot r_A \cdot r_B \cdot \sin\mu \cdot \dot\mu = 2 \cdot r_A \cdot \dot{r}_A + 2 \cdot r_B \cdot \dot{r}_B \tag{33}$$

From (33) one writes $\dot\mu$ (38), but it is necessary to obtain first $\dot{r}_A$ (34) from expression (23):

$$2 \cdot r_A \cdot \dot{r}_A = 2 \cdot r_B \cdot \dot{r}_B - 2 \cdot r_b \cdot \dot{r}_B \cdot \cos(\delta + \tau)$$
$$+ 2 \cdot r_b \cdot r_B \cdot \sin(\delta + \tau) \cdot (\dot\delta + \dot\tau) \tag{34}$$

To solve (34) we need the derivatives $\dot\delta$ and $\dot\tau$. From (7) relations (35 and 36) will be obtained. $\dot\tau$ takes the form (37):

$$\delta' = \frac{s'' \cdot (s_0 + e) - s' \cdot (s' - e)}{(s_0 + s)^2 + (s' - e)^2} \tag{35}$$

$$\dot\delta = \delta' \cdot \omega \tag{36}$$

$$\dot\tau = -\dot{\alpha}_B = -\frac{e \cdot \dot{s}}{r_B^2} \tag{37}$$

Now we can determine $\dot\mu$ (38), $\dot{\alpha}_A$ (28) and $\dot{\theta}_A$ (39):

$$\dot\mu = \frac{\dot{r}_A \cdot r_B \cdot \cos\mu + r_A \cdot \dot{r}_B \cdot \cos\mu - r_A \cdot \dot{r}_A - r_B \cdot \dot{r}_B}{r_A \cdot r_B \cdot \sin\mu} \tag{38}$$

$$\dot{\theta}_A = \dot\varphi - \dot\gamma = \omega - \dot{\alpha}_A \tag{39}$$

We write $\cos\alpha_A$ and $\sin\alpha_A$ (40-41):

$$\cos\alpha_A = \frac{e \cdot \sqrt{(s_0 + s)^2 + (s' - e)^2} + r_b \cdot (s' - e)}{r_A \cdot \sqrt{(s_0 + s)^2 + (s' - e)^2}} \tag{40}$$

$$\sin\alpha_A = \frac{(s_0 + s) \cdot \left[\sqrt{(s_0 + s)^2 + (s' - e)^2} - r_b\right]}{r_A \cdot \sqrt{(s_0 + s)^2 + (s' - e)^2}} \tag{41}$$

Further, we can obtain expression $\cos(\alpha_A - \delta)$ (42), and $\cos(\alpha_A - \delta) \cdot \cos\delta$ (43):

$$\cos(\alpha_A - \delta) = \frac{(s_0 + s) \cdot s'}{r_A \cdot \sqrt{(s_0 + s)^2 + (s' - e)^2}} = \frac{s'}{r_A} \cdot \cos\delta \tag{42}$$

$$\cos(\alpha_A - \delta) \cdot \cos\delta = \frac{s'}{r_A} \cdot \cos^2\delta \tag{43}$$

Finally the forces and the velocities are deduced as follows (48-50):

$$\begin{cases} v_a = v_m \cdot \sin(\alpha_A - \delta) \\ \\ F_a = F_m \cdot \sin(\alpha_A - \delta) \end{cases} \tag{44}$$

$$\begin{cases} v_n = v_m \cdot \cos(\alpha_A - \delta) \\ \\ F_n = F_m \cdot \cos(\alpha_A - \delta) \end{cases} \tag{45}$$

$$\begin{cases} v_i = v_n \cdot \sin\delta \\ \\ F_i = F_n \cdot \sin\delta \end{cases} \tag{46}$$

$$\begin{cases} v_2 = v_n \cdot \cos\delta = v_m \cdot \cos(\alpha_A - \delta) \cdot \cos\delta \\ \\ F_u = F_n \cdot \cos\delta = F_m \cdot \cos(\alpha_A - \delta) \cdot \cos\delta \end{cases} \tag{47}$$

2.4 Determining the efficiency of the Module B

$$P_u = F_u \cdot v_2 = F_m \cdot v_m \cdot \cos^2(\alpha_A - \delta) \cdot \cos^2\delta \tag{48}$$

$$P_c = F_m \cdot v_m \tag{49}$$

$$\begin{aligned} \eta_i = \frac{P_u}{P_c} &= \frac{F_m \cdot v_m \cdot \cos^2(\alpha_A - \delta) \cdot \cos^2\delta}{F_m \cdot v_m} = \\ &= \cos^2(\alpha_A - \delta) \cdot \cos^2\delta = [\cos(\alpha_A - \delta) \cdot \cos\delta]^2 = \\ &- [\frac{s'}{r_A} \cdot \cos^2\delta]^2 = \frac{s'^2}{r_A^2} \cdot \cos^4\delta \end{aligned} \tag{50}$$

2.5 Determining the transmission function D, for the Module B

The follower's velocity (47) can be written into the form (51):

$$v_2 = v_n \cdot \cos \delta = v_m \cdot \cos(\alpha_A - \delta) \cdot \cos \delta = v_m \cdot \frac{s'}{r_A} \cdot \cos^2 \delta =$$

(51)

$$= r_A \cdot \dot{\theta}_A \cdot \frac{s'}{r_A} \cdot \cos^2 \delta = \dot{\theta}_A \cdot s' \cdot \cos^2 \delta = \theta_A^I \cdot \omega \cdot s' \cdot \cos^2 \delta$$

With relations (51) and (52) we determine the transmission function (the dynamic modulus), D (53):

$$v_2 = s' \cdot D \cdot \omega$$

(52)

$$D = \theta_A^I \cdot \cos^2 \delta$$

(53)

Expression $\cos^2 \delta$ is known (54):

$$\cos^2 \delta = \frac{(s_0 + s)^2}{(s_0 + s)^2 + (s'-e)^2}$$

(54)

The expression of the θ'_A is more difficult (55):

$$\theta_A^I = [(s_0 + s)^2 + e^2 - e \cdot s' - r_b \cdot \sqrt{(s_0 + s)^2 + (s'-e)^2}] \cdot$$
$$\{[(s_0 + s)^2 + (s'-e)^2] \cdot \sqrt{(s_0 + s)^2 + (s'-e)^2}$$
$$+ r_b \cdot [s'' \cdot (s_0 + s) - s' \cdot (s'-e) - (s_0 + s)^2 - (s'-e)^2]\} /$$
$$[(s_0 + s)^2 + (s'-e)^2] / \{[(s_0 + s)^2 + e^2 + r_b^2] \cdot$$
$$\cdot \sqrt{(s_0 + s)^2 + (s'-e)^2} - 2 \cdot r_b \cdot [(s_0 + s)^2 + e^2 - e \cdot s']\}$$

(55)

We will determine μ by its expressions (56-57):

$$\cos \mu = \frac{[(s_0 + s)^2 + e^2] \cdot \sqrt{(s_0 + s)^2 + (s'-e)^2} - r_b \cdot [(s_0 + s)^2 + e^2 - e \cdot s']}{r_A \cdot r_B \cdot \sqrt{(s_0 + s)^2 + (s'-e)^2}}$$

(56)

$$\sin \mu = \frac{r_b \cdot (s_0 + s) \cdot s'}{r_A \cdot r_B \cdot \sqrt{(s_0 + s)^2 + (s'-e)^2}}$$

(57)

2.6 The dynamics of the Module B

For the dynamics of the Module B the relations (58-60) are used:

$$\Delta X = -\frac{\dfrac{k^2 + 2kK}{(K+k)^2} \cdot s^2 + \dfrac{2kx_0}{K+k} \cdot s + \dfrac{[\dfrac{K^2}{(K+k)^2} \cdot m_S^* + m_T^*] \cdot \omega^2}{K+k} \cdot y'^2}{2 \cdot [s + \dfrac{kx_0}{K+k}]}$$

(58)

$$\Delta X = -\frac{\dfrac{k^2 + 2kK}{(K+k)^2} \cdot s^2 + \dfrac{2kx_0}{K+k} \cdot s + \dfrac{[\dfrac{K^2}{(K+k)^2} \cdot m_s^* + m_T^*] \cdot \omega^2}{K+k} \cdot (D \cdot s')^2}{2 \cdot [s + \dfrac{kx_0}{K+k}]}$$

(59)

$$X = s + \Delta X$$

(60)

2.7 The dynamic analysis of the module B

It presents now the dynamics of the module B for some known movement laws.

We begin with the classical law SIN (see the diagram in figure 2); A speed rotation n=5500 [rot/min], for a maxim theoretical displacement of the valve h=6 [mm] is used. The phase angle is $\varphi_u=\varphi_c=65$ [degree]; the ray of the basic circle is $r_0=13$ [mm].

For the ray of the roll the value $r_b=13$ [mm] has been adopted.

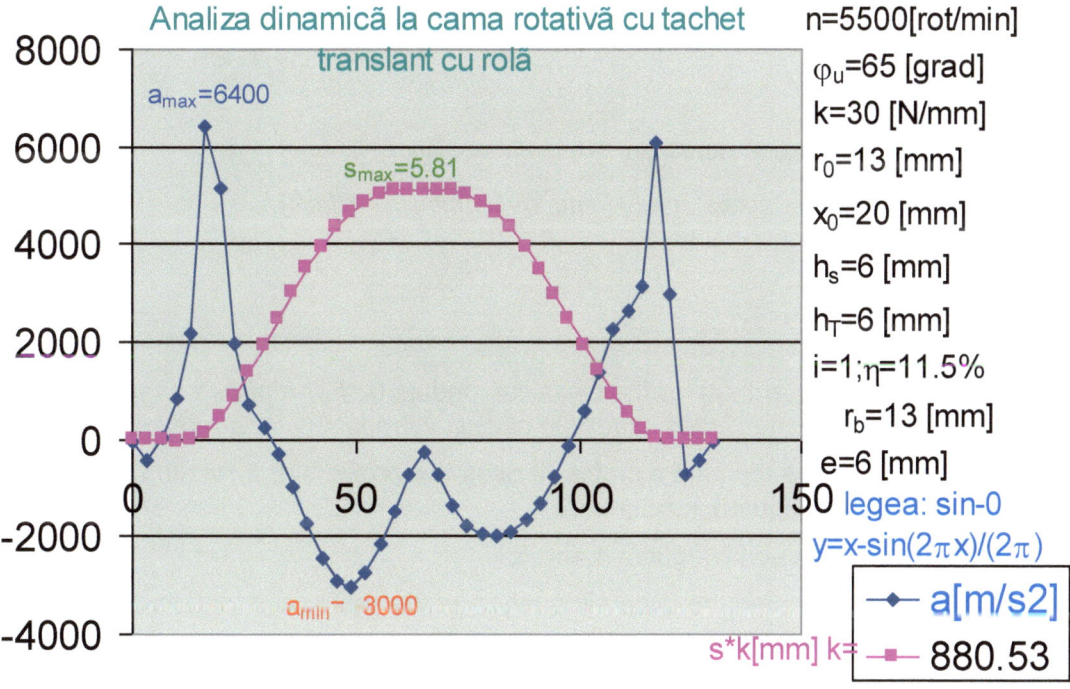

Fig. 2 *The dynamic analysis of the module B. The law SIN, n=550 rpm, $\varphi_u=65^0$, $r_0=13$ [mm], $r_b=13$ [mm], $h_T=6$ [mm], e=0 [mm], k=30 [N/mm], and $x_0=20$ [mm].*

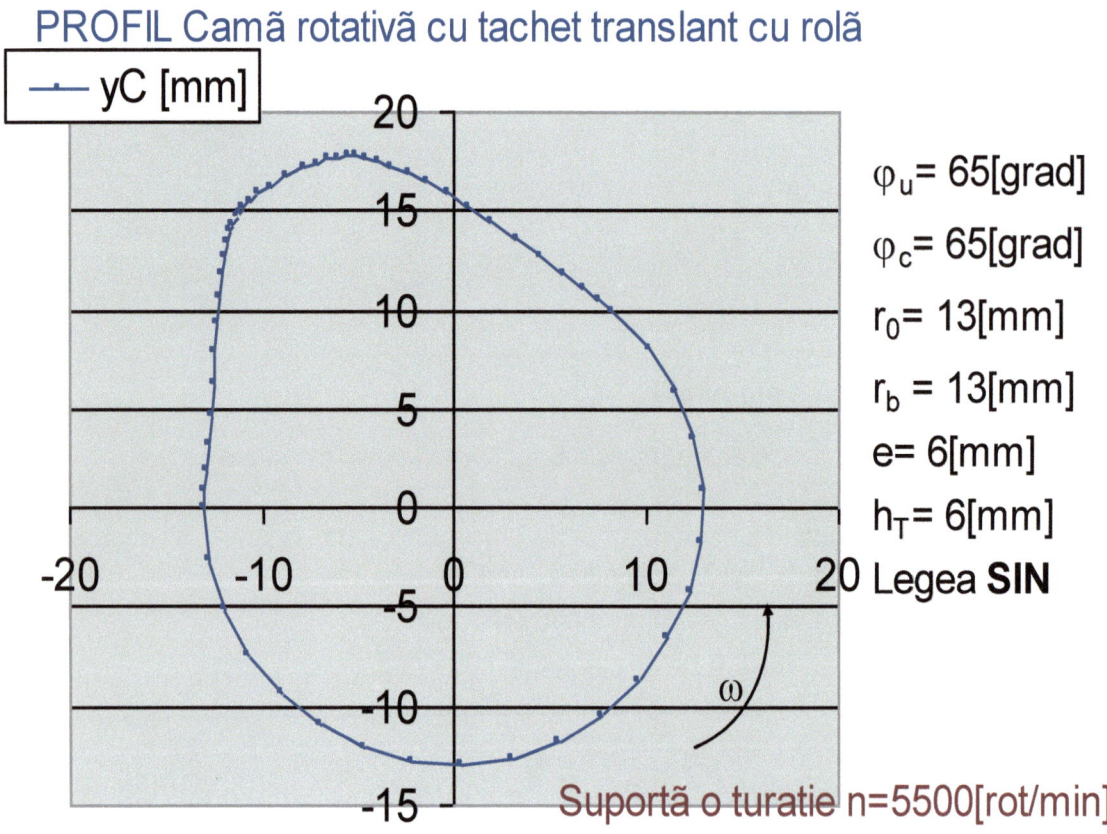

PROFIL Camă rotativă cu tachet translant cu rolă

— yC [mm]

φ_u= 65[grad]

φ_c= 65[grad]

r_0= 13[mm]

r_b = 13[mm]

e= 6[mm]

h_T= 6[mm]

Legea **SIN**

ω

Suportă o turatie n=5500[rot/min]

Fig. 3 *The profile SIN at the module B. n=5500 rpm*
φ_u=65^0, r_0=13 [mm], r_b=13 [mm], h_T=6 [mm].

The dynamics are better than for the classical module C. *For a phase angle of just 65 degrees the accelerations have the same values as for the classical module C for a relaxed phase (75^0-80^0).*

In figure 3 we can see the cam's profile. It uses the profile sin, a rotation speed n=5500 rpm, and φ_u=65^0, r_0=13 [mm], r_b=13 [mm], h_T=6 [mm].

The law COS can be seen in figures 4 and 5.

In the figure 4 is presented the dynamic analyze of the profile cos, and its profile design can be seen in the figure 5.

The principal parameters are:

Law COS, n=5500 rpm, φ_u=65^0, r_0=13 [mm], r_b=6 [mm], h_T=6 [mm], η=10.5%.

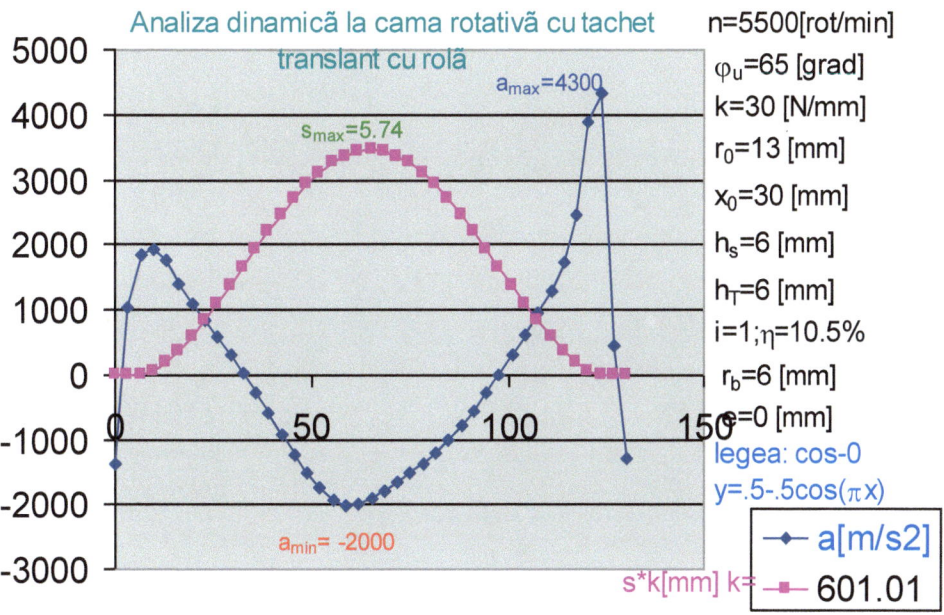

Fig. 4 *The dynamic analysis of the module B. Law COS, n=5500 rpm, φ_u=65⁰, r_0=13 [mm], r_b=6 [mm], h_T=6 [mm], η=10.5%.*

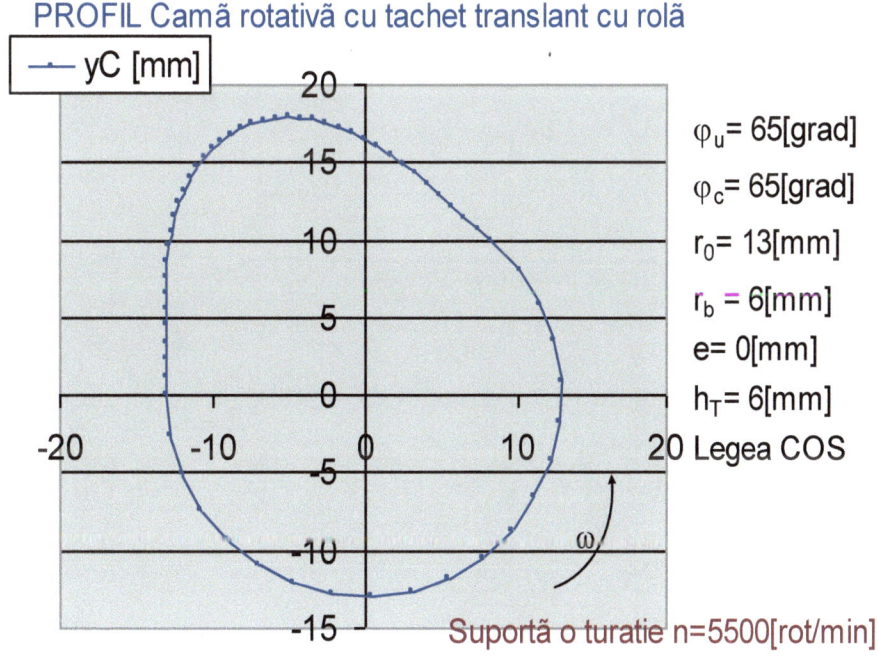

Fig. 5 *The profile COS at the module B, n=5500 rpm, φ_u=65⁰, r_0=13 [mm], r_b=6 [mm], h_T=6 [mm].*

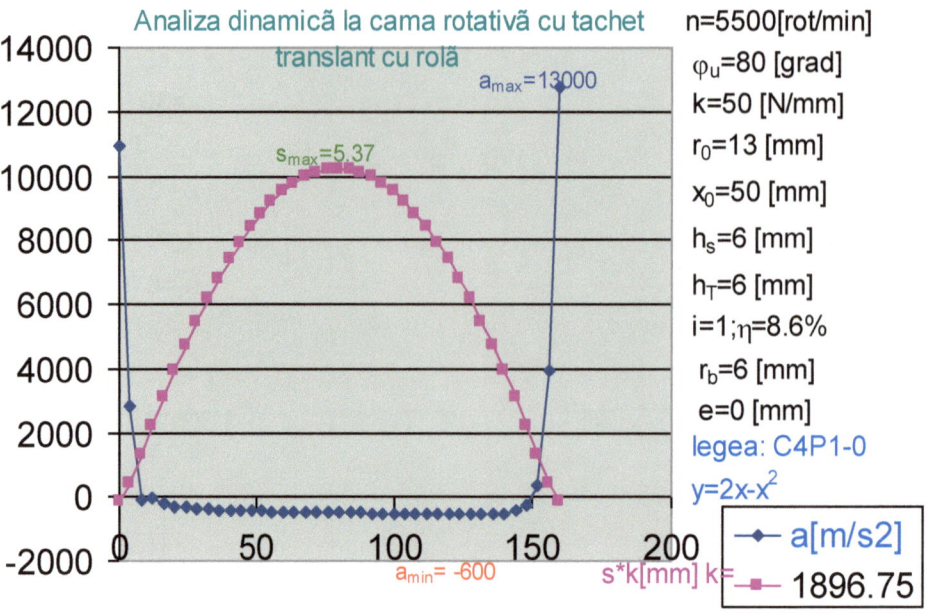

Fig. 6 *The dynamic analyze. Law C4P1-0, n=5500 rpm, φ_u=80⁰, r_0=13 [mm], r_b=6 [mm], h_T=6 [mm].*

In figure 6 the law C4P, created by the authors, is analyzed dynamic. The vibrations are diminished, the noises are limited, the effective displacement of the valve is increased, s_{max}=5.37 [mm].

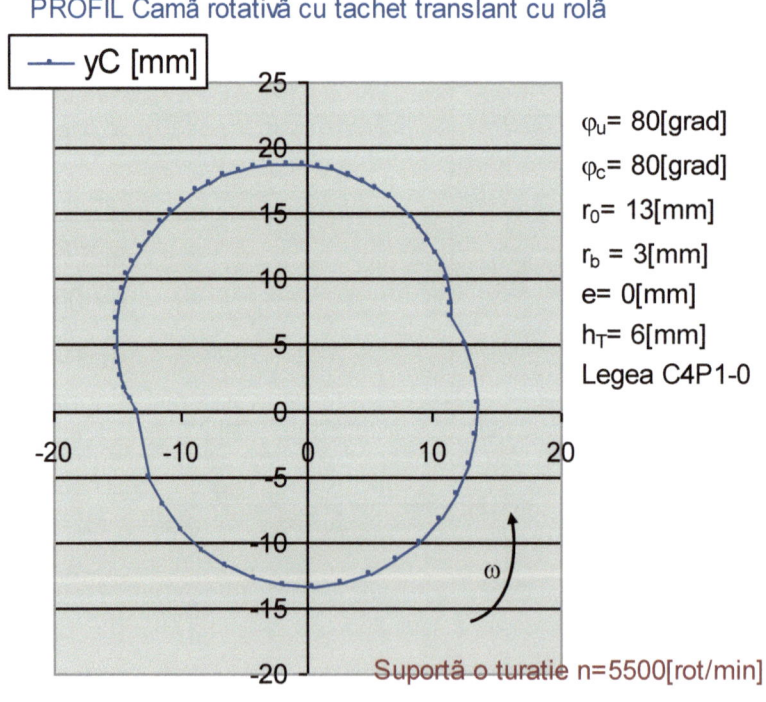

Fig. 7 *The profile C4P of the module B.*

The efficiency has a good value η=8.6%. In figure 7 the profile of C4P law is presented. It starts at the law C4P with n=5500 [rpm], but for this law the rotation velocity can increase to high values of 30000-40000 [rpm] (see Fig. 8).

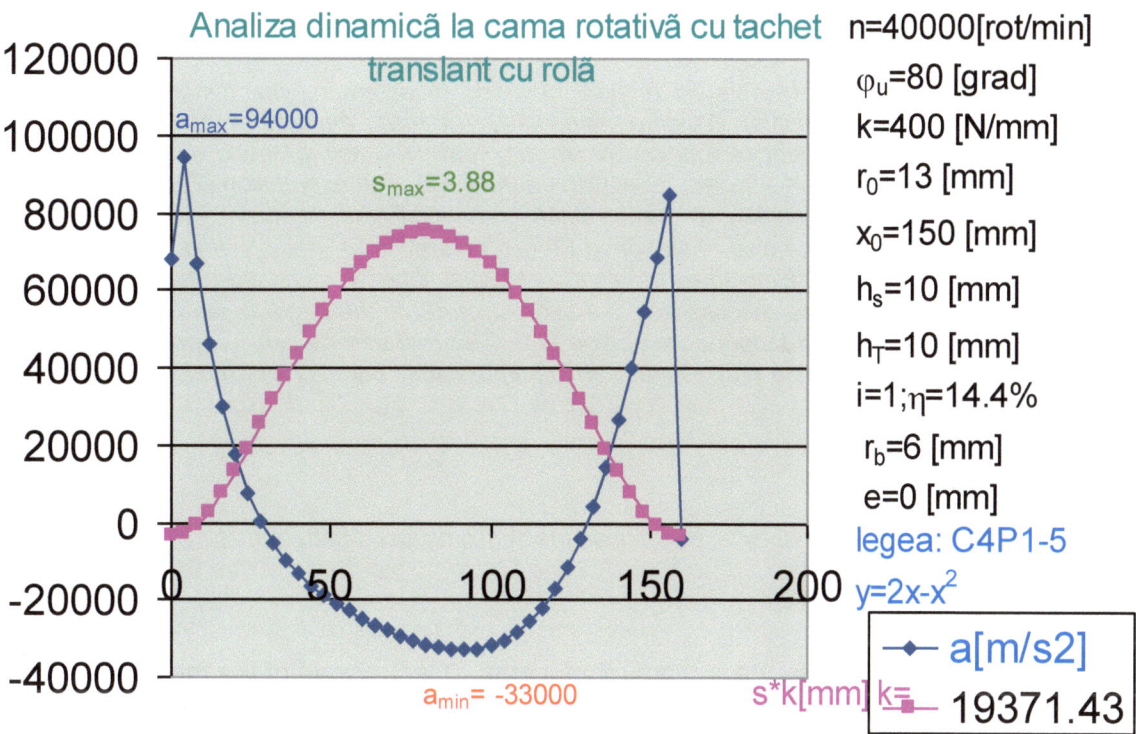

Fig. 8 *The dynamic analysis of the module B. Law C4P1-5, n=40000 rpm.*

3 Conclusions

We can speak about an advantage of the module B in comparison to the classical module C. With the module B, (when the follower is provided with a roll) it can obtain high rotation velocity with superior efficiency.

References

[1] Petrescu F.I., Petrescu R.V., *Contributions at the dynamics of cams.* In the Ninth IFToMM International Sympozium on Theory of Machines and Mechanisms, SYROM 2005, Bucharest, Romania, Vol. I, pp. 123-128, 2005.

CHAPTER V
DYNAMICS OF THE CLASSIC DISTRIBUTION

Abstract: *This chapter presents an original methods to determine the dynamic parameters at the camshaft (the distribution mechanisms). We determine initially the mass moment of inertia (mechanical) of the mechanism, reduced to the element of rotation, ie at cam (basically using kinetic energy conservation, the system 1). Average moment of inertia is calculated with equation (2). The expression (2) depends on the type of cam-tappet mechanism, and of the law of motion used both uphill and downhill. The angular velocity is a function of the position cam (φ) but also of its speed (3). To determine ω^2 (relationship 3) have found J *, and more specifically Jmax. Differentiating the formula (6), against time, is obtained the angular acceleration expression (8). Differentiating twice successively, the expression (9) in the angle φ, we obtain a reduced tappet speed (equation 10), and reduced tappet acceleration (11). The real and dynamic, tappet acceleration can be determined directly using the relation (12). General (original) dynamic equations of motion for the determination of ω and ε have the form (13).*

Keywords: *cam, cams, cam mechanisms, distribution mechanisms, camshaft, tappet.*

We determine initially the mass moment of inertia (mechanical) of the mechanism, reduced to the element of rotation, ie at cam (basically using kinetic energy conservation, the system 1).

$$\begin{cases} J_{cama} = \frac{1}{2} \cdot M_c \cdot R^2 \\[2mm] R^2 = (R_0 + s)^2 + s'^2 \\[2mm] J_{cama} = \frac{1}{2} \cdot M_c \cdot \left[(R_0 + s)^2 + s'^2\right] \\[2mm] J^* = \frac{1}{2} \cdot M_c \cdot \left[(R_0 + s)^2 + s'^2\right] + m_T \cdot s'^2 \\[2mm] J^* = \frac{1}{2} \cdot M_c \cdot R_0^2 + \frac{1}{2} \cdot M_c \cdot s^2 + M_c \cdot R_0 \cdot s + \frac{1}{2} \cdot M_c \cdot s'^2 + m_T \cdot s'^2 \\[2mm] J^* = J_{constant} + J \\[2mm] J \equiv J_{variabil} = \frac{1}{2} \cdot M_c \cdot s^2 + M_c \cdot R_0 \cdot s + \frac{1}{2} \cdot M_c \cdot s'^2 + m_T \cdot s'^2 \end{cases} \qquad (1)$$

Average moment of inertia is calculated with equation (2).

$$J_m^* = \frac{J_{min}^* + J_{max}^*}{2} = \frac{1}{2} \cdot M_c \cdot R_0^2 + \frac{J_{max}}{2} \qquad (2)$$

The expression (2) depends on the type of cam-tappet mechanism, and of the law of motion used both uphill and downhill. The angular velocity is a function of the position cam (φ) but also of its speed (3).

$$\omega^2 = \frac{J_m^* \cdot \omega_m^2}{J^*} \tag{3}$$

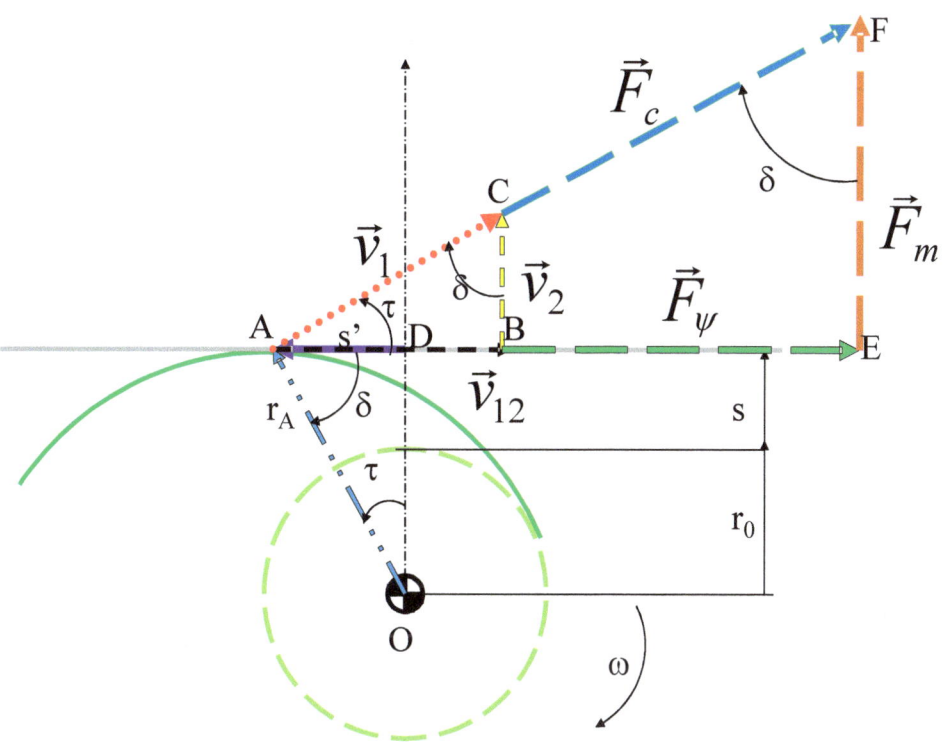

Fig. 1 *Forces and speeds to the cam with plate translated follower*

To determine ω^2 (relationship 3) have found J *, and more specifically Jmax.

And at the classic distribution (rotative cam, and plat tappet in translational motion), the relationship which determine the Jmax, depend and of the law of movement.

We start the simulation with a classical law of motion, namely the cosine law. At the climb, the cosine law is expressed by relations (4).

$$\begin{cases} s = \dfrac{h}{2} - \dfrac{h}{2} \cdot \cos\left(\pi \cdot \dfrac{\varphi}{\varphi_u}\right) \\[2ex] s' \equiv v_r = \dfrac{\pi \cdot h}{2 \cdot \varphi_u} \cdot \sin\left(\pi \cdot \dfrac{\varphi}{\varphi_u}\right) \\[2ex] s'' \equiv a_r = \dfrac{\pi^2 \cdot h}{2 \cdot \varphi_u^2} \cdot \cos\left(\pi \cdot \dfrac{\varphi}{\varphi_u}\right) \\[2ex] s''' \equiv \alpha_r = -\dfrac{\pi^3 \cdot h}{2 \cdot \varphi_u^3} \cdot \sin\left(\pi \cdot \dfrac{\varphi}{\varphi_u}\right) \end{cases} \tag{4}$$

Where φ varies from 0 to φ_u. It achieves J_{max} for $\varphi=\varphi_u/2$.

$$J_{max} = M_c \cdot \left[\frac{h^2}{8} + R_0 \cdot \frac{h}{2} + \frac{1}{8} \cdot \frac{\pi^2 \cdot h^2}{\varphi_u^2} \right] + m_T \cdot \frac{\pi^2 \cdot h^2}{4 \cdot \varphi_u^2} \tag{5}$$

The expression (3) now takes the form (6).

$$\begin{cases} \omega^2 = \omega_m^2 \cdot \dfrac{A}{B} \\[2mm] A = M_c \cdot R_0^2 + M_c \cdot \dfrac{h^2}{8} + \dfrac{1}{2} \cdot M_c \cdot R_0 \cdot h + \\[2mm] + \dfrac{1}{8} \cdot M_c \cdot \dfrac{\pi^2 \cdot h^2}{\varphi_u^2} + \dfrac{1}{4} \cdot m_T \cdot \dfrac{\pi^2 \cdot h^2}{\varphi_u^2} \\[2mm] B = M_c \cdot R_0^2 + M_c \cdot s^2 + 2 \cdot M_c \cdot R_0 \cdot s + M_c \cdot s'^2 + 2 \cdot m_T \cdot s'^2 \\[2mm] \omega = \omega_m \cdot \sqrt{\dfrac{A}{B}} \end{cases} \tag{6}$$

Where ω_m is the nominal cam velocity and it express at the distribution mechanisms, based on the speed shaft, with relationship (7).

$$\omega_m = 2 \cdot \pi \cdot v_c = 2 \cdot \pi \cdot \frac{n_c}{60} = \frac{2 \cdot \pi}{60} \cdot \frac{n_{motor}}{2} = \frac{\pi \cdot n}{60} \tag{7}$$

Differentiating the formula (6), against time, is obtained the angular acceleration expression (8).

$$\varepsilon = -\omega^2 \cdot \frac{(M_c \cdot s + M_c \cdot R_0 + M_c \cdot s'' + 2 \cdot m_T \cdot s'') \cdot s'}{B} \tag{8}$$

For a classic cam and tappet mechanism (without valve) dynamic movement tappet is expressed by equation (9), who was presented and derived in Chapter 2 (equation 48), and now by canceling valve mass, will customize and reaching form below (9).

$$x = s - \frac{(K+k) \cdot m_T \cdot \omega^2 \cdot s'^2 + (k^2 + 2k \cdot K) \cdot s^2 + 2k \cdot x_0 \cdot (K+k) \cdot s}{2 \cdot (K+k)^2 \cdot \left(s + \dfrac{k \cdot x_0}{K+k} \right)} \tag{9}$$

Where x is the dynamic movement of the pusher, while s is its normal, kinematics movement. K is the spring constant of the system, and k is the spring constant of the tappet spring. It note, with x_0 the tappet spring preload, with m_T the mass of the tappet, with ω the angular rotation speed of the cam (or camshaft), where s' is the first derivative in function of φ of the tappet movement, s. Differentiating twice successively, the expression (9) in the angle φ, we obtain a reduced tappet speed (equation 10), and reduced tappet acceleration (11).

$$\begin{cases} N = (K+k) \cdot m_T \cdot \omega^2 \cdot s'^2 + (k^2 + 2k \cdot K) \cdot s^2 + 2k \cdot x_0 \cdot (K+k) \cdot s \\[2mm] M = \left[(K+k) m_T \omega^2 \cdot 2s's'' + \left(k^2 + 2kK \right) \cdot 2ss' + 2kx_0 (K+k) \cdot s' \right] \cdot \\[2mm] \quad \cdot \left(s + \dfrac{kx_0}{K+k} \right) - N \cdot s' \\[4mm] x' = s' - \dfrac{M}{2 \cdot (K+k)^2 \cdot \left(s + \dfrac{kx_0}{K+k} \right)^2} \end{cases} \tag{10}$$

$$\begin{cases} N = (K+k) \cdot m_T \cdot \omega^2 \cdot s'^2 + (k^2 + 2k \cdot K) \cdot s^2 + 2k \cdot x_0 \cdot (K+k) \cdot s \\[3mm] M = \left[(K+k) m_T \omega^2 \cdot 2s's'' + \left(k^2 + 2kK \right) \cdot 2ss' + 2kx_0 (K+k) \cdot s' \right] \cdot \\[2mm] \quad \cdot \left(s + \dfrac{kx_0}{K+k} \right) - N \cdot s' \\[4mm] O = (K+k) \cdot m_T \cdot \omega^2 \cdot 2 \cdot \left(s''^2 + s' \cdot s''' \right) + \\[2mm] \quad + \left(k^2 + 2 \cdot k \cdot K \right) \cdot 2 \cdot \left(s'^2 + s \cdot s'' \right) + 2 \cdot k \cdot x_0 \cdot (K+k) \cdot s'' \\[4mm] x'' = s'' - \dfrac{\left[O \cdot \left(s + \dfrac{kx_0}{K+k} \right) - N \cdot s'' \right] \cdot \left(s + \dfrac{kx_0}{K+k} \right) - M \cdot 2 \cdot s'}{2 \cdot (K+k)^2 \cdot \left(s + \dfrac{kx_0}{K+k} \right)^3} \end{cases} \tag{11}$$

The real and dynamic, tappet acceleration can be determined directly using the relation (12).

$$\ddot{x} = x'' \cdot \omega^2 + x' \cdot \varepsilon \tag{12}$$

General (original) dynamic equations of motion for the determination of ω and ε have the form (13).

$$\begin{cases} \omega^2 = \dfrac{J_m^*}{J^*} \cdot \omega_m^2; \quad \omega = \sqrt{\dfrac{J_m^*}{J^*}} \cdot \omega_m \\[4mm] \varepsilon = -\dfrac{1}{2} \cdot \omega^2 \cdot \dfrac{J^{*'}}{J^*} \end{cases} \tag{13}$$

With a program (written in excel) one obtains the diagrams of the movement laws (see the Figure 2), the dynamic tappet acceleration for a n=5500 [rpm] (Fig. 3), and the cam profile (Fig. 4).

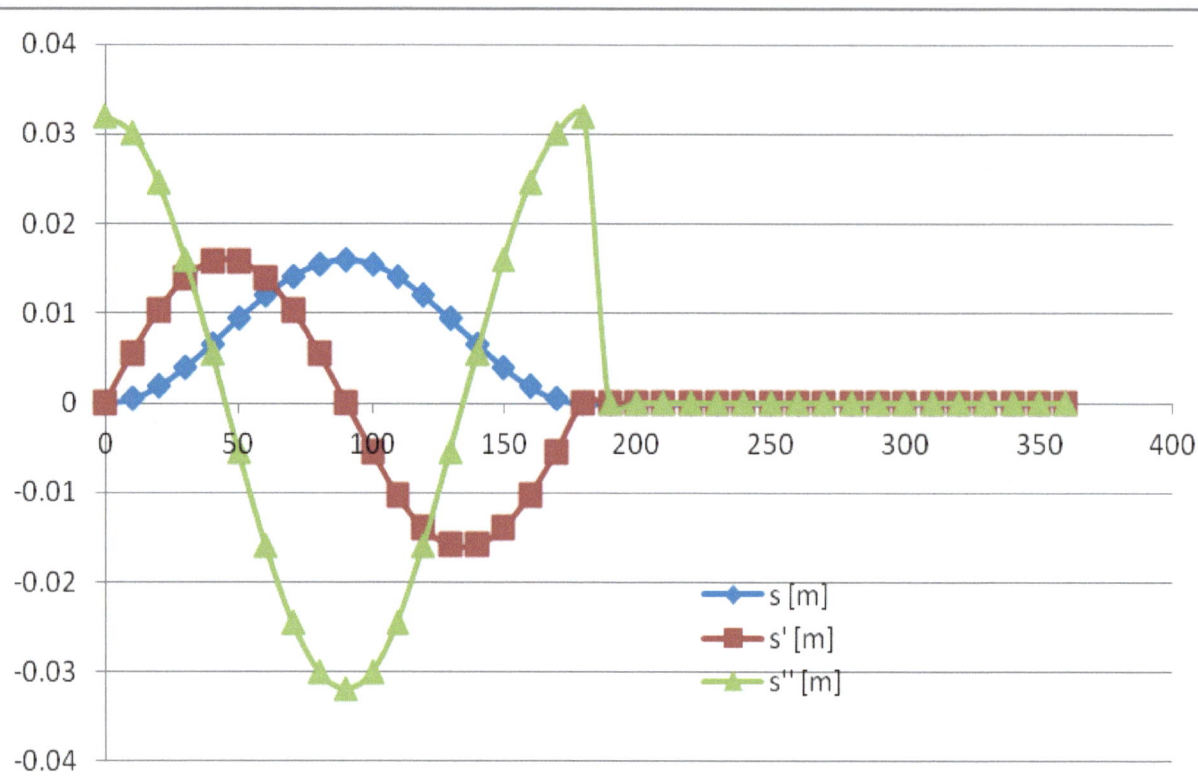

Fig. 2 *s, s', s'' diagrams at the cam with plate translated follower*

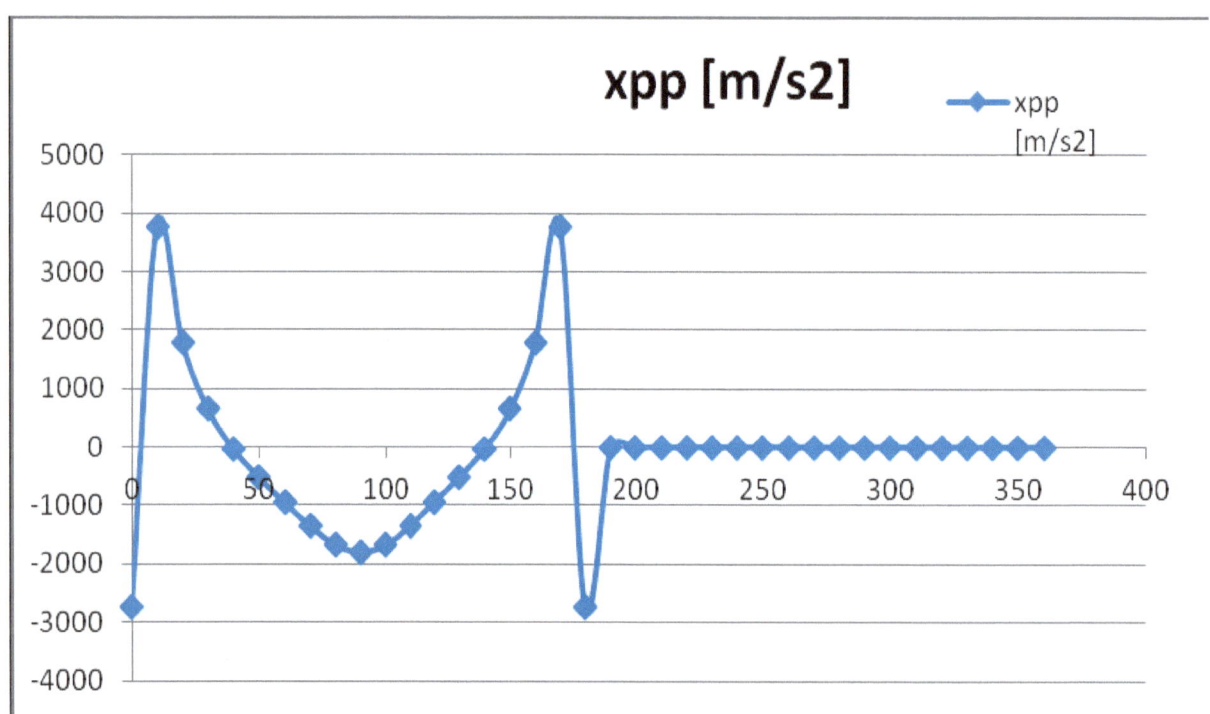

Fig. 3 *The dynamic tappet acceleration at the cam with plate translated follower*

The profile synthesis was made with the system of relations (14) when the cam is moving in the orar sense, and (15) when the cam is rotating trigonometric.

$$\begin{cases} x_c = -s' \cdot \cos\varphi - (r_0 + s) \cdot \sin\varphi \\ y_c = (r_0 + s) \cdot \cos\varphi - s' \cdot \sin\varphi \end{cases} \qquad (14)$$

$$\begin{cases} x_c = s' \cdot \cos\varphi + (r_0 + s) \cdot \sin\varphi \\ y_c = (r_0 + s) \cdot \cos\varphi - s' \cdot \sin\varphi \end{cases} \qquad (15)$$

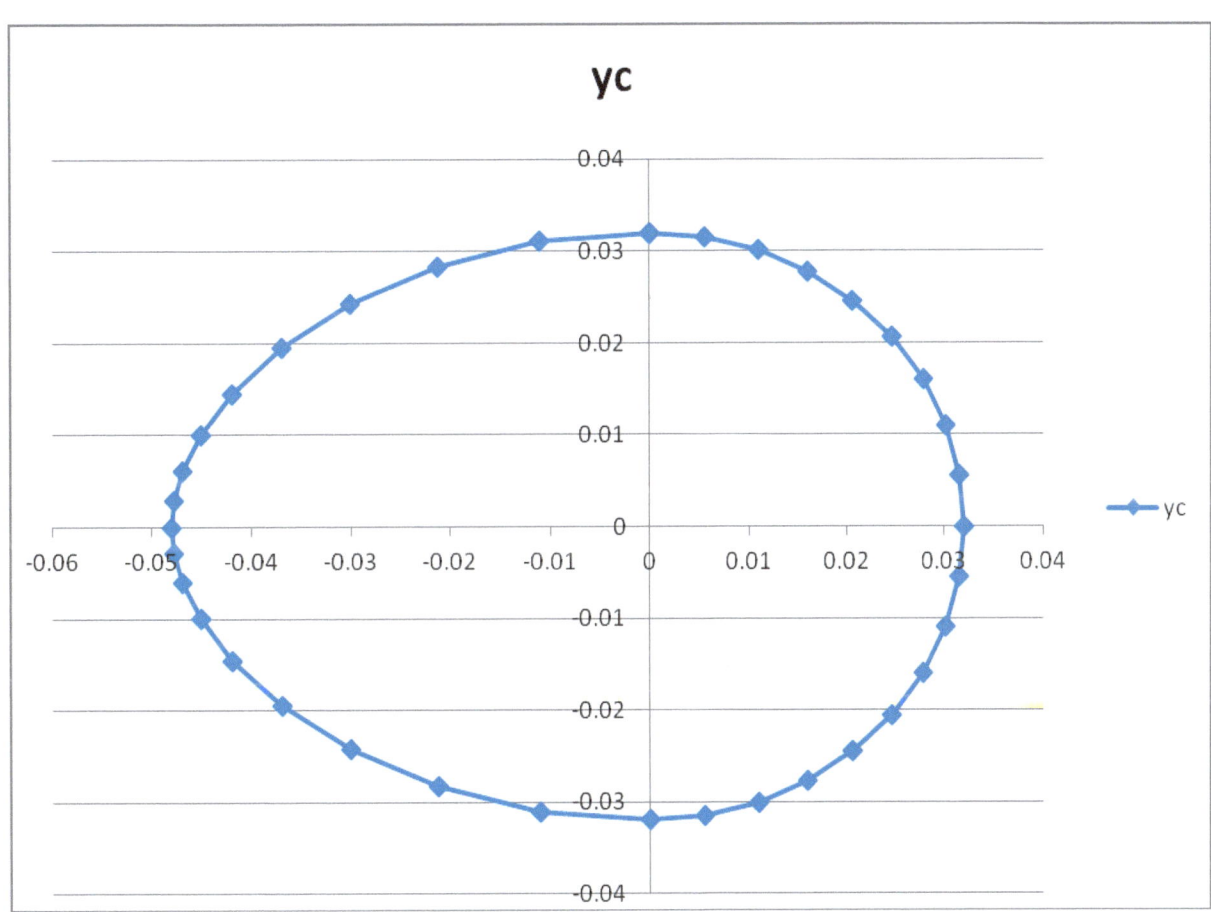

Fig. 4 *The cam profile, at the cam with plate translated follower*

CHAPTER VI
PRECISION OF THE CLASSIC DISTRIBUTION

Abstract: *This chapter presents an original methods to determine the dynamic parameters at the camshaft (the distribution mechanisms), when the cam was made to work normally. We can make the geometrical synthesis of the cam profile with the help of the cinematics of the mechanism. One uses as well the reduced speed, s'. The reduced velocity, s', folded to 90 degrees, completes the triangle OAB. We can determine the coordinates of the point A from the tappet (1), and from the cam (2). The forces and the velocities at a cam with plate translated tappet can be seen in the figure 2. The driving force Fm, perpendicular on the r in A, is decomposed in two forces: the utile force Fu, which acts the tappet and the lost force Fa, who is a slipping force. The velocities take the same positions (system 4). The efficiency of the mechanism is determining with the relationship (5). We can make the geometro-kinematics synthesis of the cam profile with the help of the cinematics of the mechanism (see the Figure 3). Now, we can make the geometro-kinematics synthesis of the classic cam profile (system 7). The moments of inertia is determined with the relationships from the system 8. The angular velocity ω, and the angular acceleration, are determined with the presented relationships (system 9). For a classic cam and tappet mechanism (without valve) dynamic movement tappet is expressed by equation (10), who was presented and derived in Chapter 2 (equation 48), and now by canceling valve mass, will customize and reaching form below (10). Where x is the dynamic movement of the pusher, while s is its normal, kinematics movement. K is the spring constant of the system, and k is the spring constant of the tappet spring. It note, with x_0 the tappet spring preload, with m_T the mass of the tappet, with ω the angular rotation speed of the cam (or camshaft), where s' is the first derivative in function of φ of the tappet movement, s. Differentiating twice successively, the expression (10) in the angle φ, we obtain a reduced tappet speed (equation 11), and reduced tappet acceleration (12). The real and dynamic, tappet acceleration can be determined directly using the relation (13). The presented dynamic system has the advantage to has a normal functionality. The synthesis was made using the natural geometro-kinematics parameters (of cam mechanism).*

Keywords: *cam, cams, cam mechanisms, distribution mechanisms, camshaft, tappet.*

1. Geometrical synthesis of the cam profile

We can make the geometrical synthesis of the cam profile with the help of the cinematics of the mechanism. One uses as well the reduced speed, s'. The reduced velocity, s', folded to 90 degrees, completes the triangle OAB (see the Figure 1).

$OB = r_0 + s$; $BA = s'$; $OA = r = r_A$; $r^2 = r_A^2 = (r_0 + s)^2 + s'^2$

It establishes a system fixed Cartesian, $xOy = x_f Oy_f$, and a mobil Cartesian system, $xOy = x_m Oy_m$ fixed with the cam.

From the lower position 0, the tappet, pushed by cam, uplifts to a general position, when the cam rotates with the φ angle. The contact point A, go from A_i^0 to A^0 (on the cam), and to A (on the tappet). The position angle of the point A from the tappet is θ_f, and from the cam is θ_m. We can determine the coordinates of the point A from the tappet (1), and from the cam (2).

$$\begin{cases} x_T \equiv x_A^f = s' = r_A \cdot \cos\theta_f \\ y_T \equiv y_A^f = r_0 + s = r_A \cdot \sin\theta_f \end{cases} \quad (1)$$

$$\begin{cases} x_c \equiv x_A^m = r_A \cdot \cos\theta_m = r \cdot \cos(\theta_f - \varphi) = r\cos\theta_f \cos\varphi + r\sin\theta_f \sin\varphi = \\ = x_T \cos\varphi + y_T \sin\varphi = s' \cdot \cos\varphi + (r_0 + s) \cdot \sin\varphi \\ \\ y_c \equiv y_A^m = r_A \cdot \sin\theta_m = r \cdot \sin(\theta_f - \varphi) = r\sin\theta_f \cos\varphi - r\sin\varphi\cos\theta_f = \\ = y_T \cos\varphi - x_T \sin\varphi = (r_0 + s) \cdot \cos\varphi - s' \cdot \sin\varphi \end{cases} \qquad (2)$$

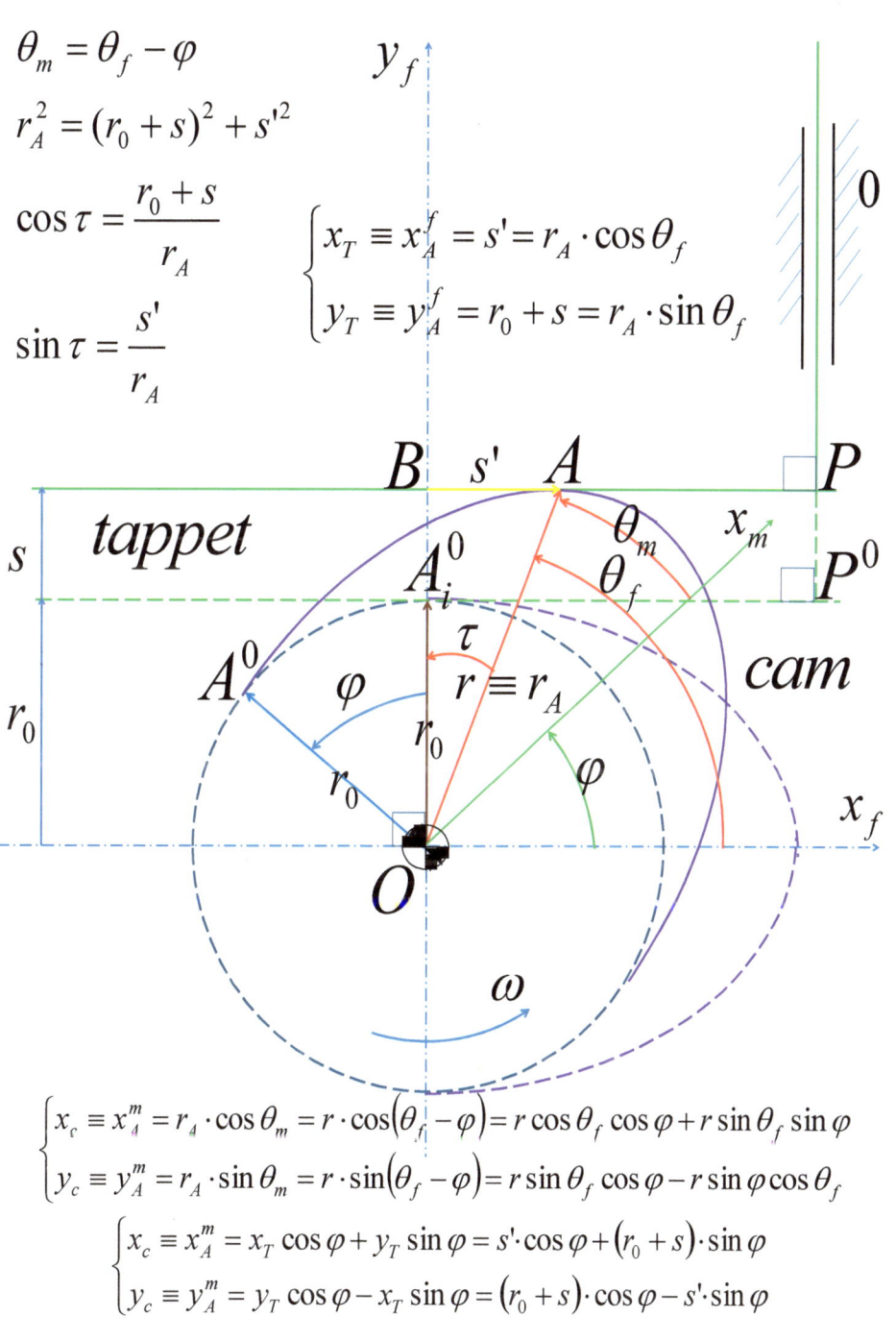

$$\theta_m = \theta_f - \varphi$$
$$r_A^2 = (r_0 + s)^2 + s'^2$$
$$\cos\tau = \frac{r_0 + s}{r_A}$$
$$\sin\tau = \frac{s'}{r_A}$$

$$\begin{cases} x_T \equiv x_A^f = s' = r_A \cdot \cos\theta_f \\ y_T \equiv y_A^f = r_0 + s = r_A \cdot \sin\theta_f \end{cases}$$

$$\begin{cases} x_c \equiv x_A^m = r_A \cdot \cos\theta_m = r \cdot \cos(\theta_f - \varphi) = r\cos\theta_f \cos\varphi + r\sin\theta_f \sin\varphi \\ y_c \equiv y_A^m = r_A \cdot \sin\theta_m = r \cdot \sin(\theta_f - \varphi) = r\sin\theta_f \cos\varphi - r\sin\varphi\cos\theta_f \end{cases}$$

$$\begin{cases} x_c \equiv x_A^m = x_T \cos\varphi + y_T \sin\varphi = s' \cdot \cos\varphi + (r_0 + s) \cdot \sin\varphi \\ y_c \equiv y_A^m = y_T \cos\varphi - x_T \sin\varphi = (r_0 + s) \cdot \cos\varphi - s' \cdot \sin\varphi \end{cases}$$

Fig. 1 *Geometry of the cam with plate translated follower*

One uses the relationships (3).

$$\begin{cases} \theta_m = \theta_f - \varphi \\ r_A^2 = (r_0 + s)^2 + s'^2 \\ \cos \tau = \dfrac{r_0 + s}{r_A} \\ \sin \tau = \dfrac{s'}{r_A} \end{cases} \qquad (3)$$

Now, we shall see the forces, the powers and the efficiency.

2. The efficiency of the cam

The forces and the velocities at a cam with plate translated tappet can be seen in the figure 2.

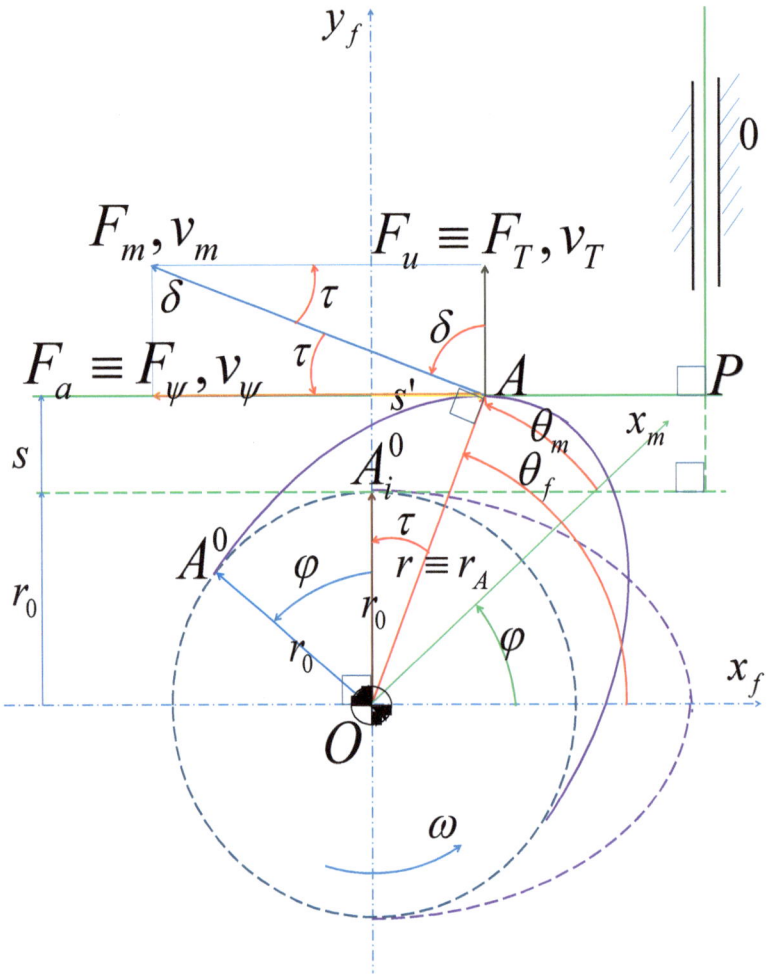

Fig. 2 *Forces and velocities of the cam with plate translated follower*

The driving force Fm, perpendicular on the r in A, is decomposed in two forces: the utile force Fu, which acts the tappet and the lost force Fa, who is a slipping force. The velocities take the same positions (system 4).

$$\begin{cases} F_u = F_m \cdot \sin\tau; & F_a = F_m \cdot \cos\tau \\ v_u = v_m \cdot \sin\tau; & v_a = v_m \cdot \cos\tau \end{cases} \tag{4}$$

The efficiency of the mechanism is determining with the relationship (5).

$$\eta_i = \frac{P_u}{P_c} = \frac{F_u \cdot v_u}{F_m \cdot v_m} = \frac{F_m \cdot \sin\tau \cdot v_m \cdot \sin\tau}{F_m \cdot v_m} = \sin^2\tau = \cos^2\delta \tag{5}$$

3. Geometro-kinematics synthesis of the cam profile

We can make the geometro-kinematics synthesis of the cam profile with the help of the cinematics of the mechanism (see the Figure 3).

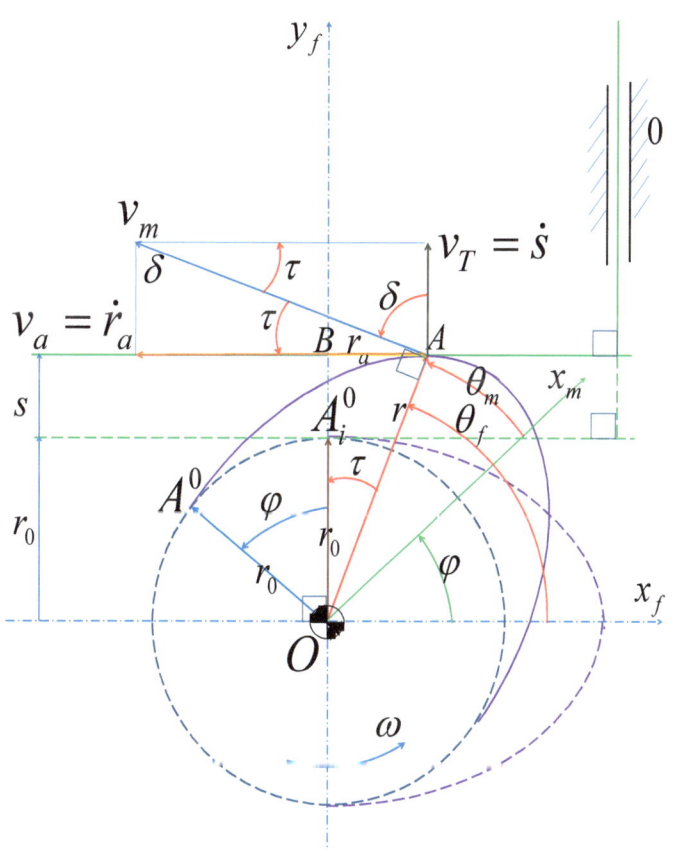

Fig. 3 *Geometro-kinematics synthesis of the cam with plate translated follower*

We denote the distance BA with r_a. We can write the relationships (6).

$$\tan \tau = \begin{cases} \cfrac{r_a}{r_0 + s} \\ \cfrac{\dot{s}}{\dot{r}_a} = \cfrac{\frac{ds}{dt}}{\frac{dr_a}{dt}} = \cfrac{ds}{dr_a} \end{cases} \Rightarrow \cfrac{r_a}{r_0 + s} = \cfrac{ds}{dr_a} \Rightarrow r_a dr_a = (r_0 + s) ds \Rightarrow$$

$$\Rightarrow \frac{1}{2} r_a^2 = r_0 \cdot s + \frac{1}{2} s^2 \Rightarrow r_a^2 = 2 \cdot r_0 \cdot s + s^2 \Rightarrow r_a = \sqrt{2 \cdot r_0 \cdot s + s^2}$$

$$\begin{cases} r = \sqrt{r_0^2 + 4 \cdot r_0 \cdot s + 2 \cdot s^2} \\[2mm] \cos \tau = \cfrac{r_0 + s}{\sqrt{r_0^2 + 4 \cdot r_0 \cdot s + 2 \cdot s^2}} \\[2mm] \sin \tau = \cfrac{\sqrt{2 \cdot r_0 \cdot s + s^2}}{\sqrt{r_0^2 + 4 \cdot r_0 \cdot s + 2 \cdot s^2}} \\[2mm] tg \tau = \cfrac{\sqrt{2 \cdot r_0 \cdot s + s^2}}{r_0 + s} \end{cases} \qquad (6)$$

Now, we can make the geometro-kinematics synthesis of the classic cam profile (system 7).

$$\begin{cases} \begin{cases} x_T = r \cdot \cos \theta_f = r_a = \sqrt{2 \cdot r_0 \cdot s + s^2} \\ y_T = r \cdot \sin \theta_f = r_0 + s \end{cases} \\[6mm] \begin{cases} x_c = r \cdot \cos \theta_m = r \cdot \cos(\theta_f - \varphi) = r \cos \theta_f \cos \varphi + r \sin \theta_f \sin \varphi = x_T \cos \varphi + y_T \sin \varphi = \\ = r_a \cos \varphi + (r_0 + s) \sin \varphi = \sqrt{2 \cdot r_0 \cdot s + s^2} \cdot \cos \varphi + (r_0 + s) \cdot \sin \varphi \\[4mm] y_c = r \cdot \sin \theta_m = r \cdot \sin(\theta_f - \varphi) = r \sin \theta_f \cos \varphi - r \sin \varphi \cos \theta_f = y_T \cos \varphi - x_T \sin \varphi = \\ = (r_0 + s) \cos \varphi - r_a \sin \varphi = (r_0 + s) \cdot \cos \varphi - \sqrt{2 \cdot r_0 \cdot s + s^2} \cdot \sin \varphi \end{cases} \end{cases} \qquad (7)$$

For a law cos, the profile takes the below form, bean-shaped (see the Figure 4).

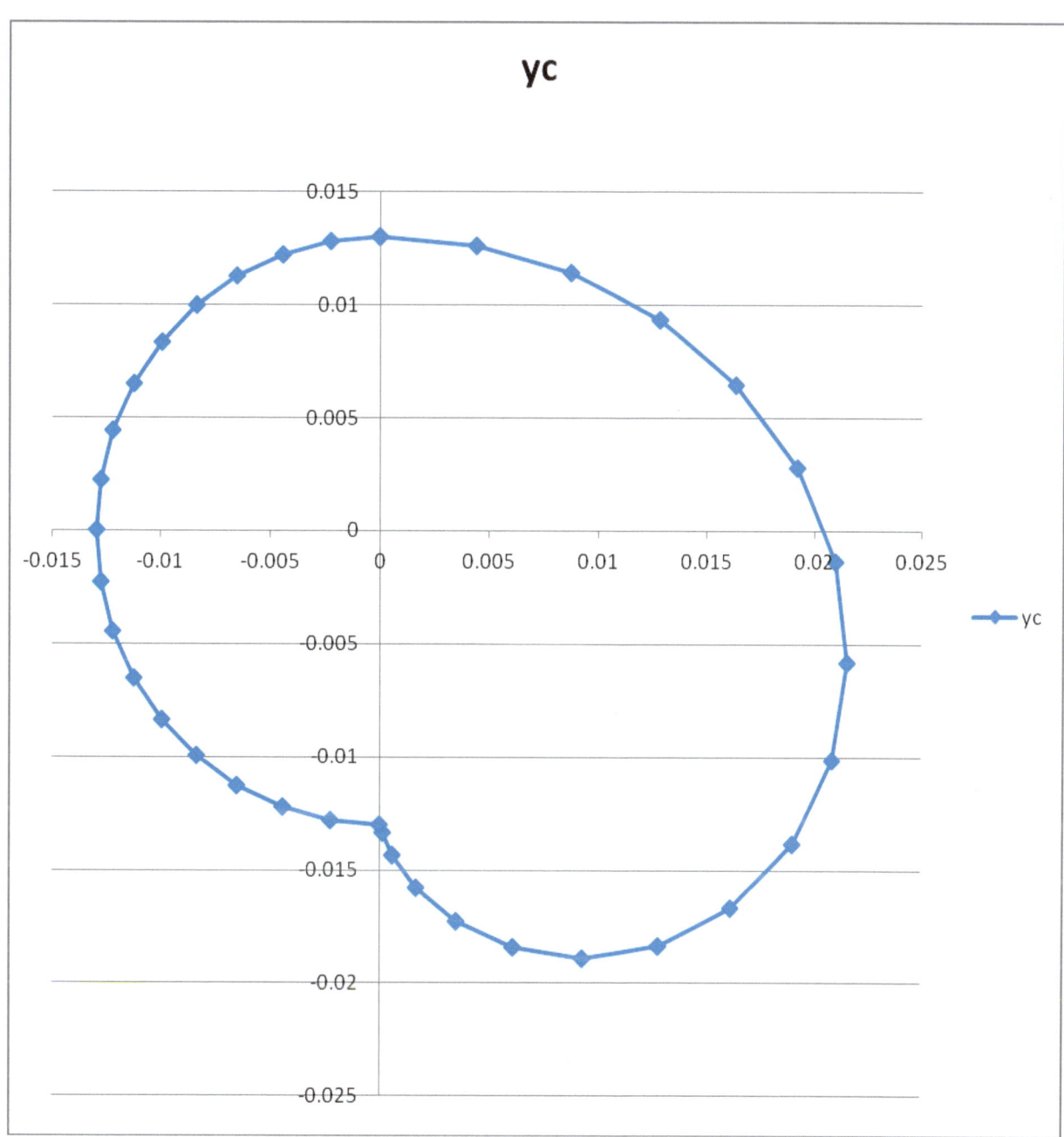

Fig. 4 *The profile of cam (bean-shaped) to the cam with plate translated follower*

This profile can be closed with an additional curve, and one obtains the form in the Figure 5.

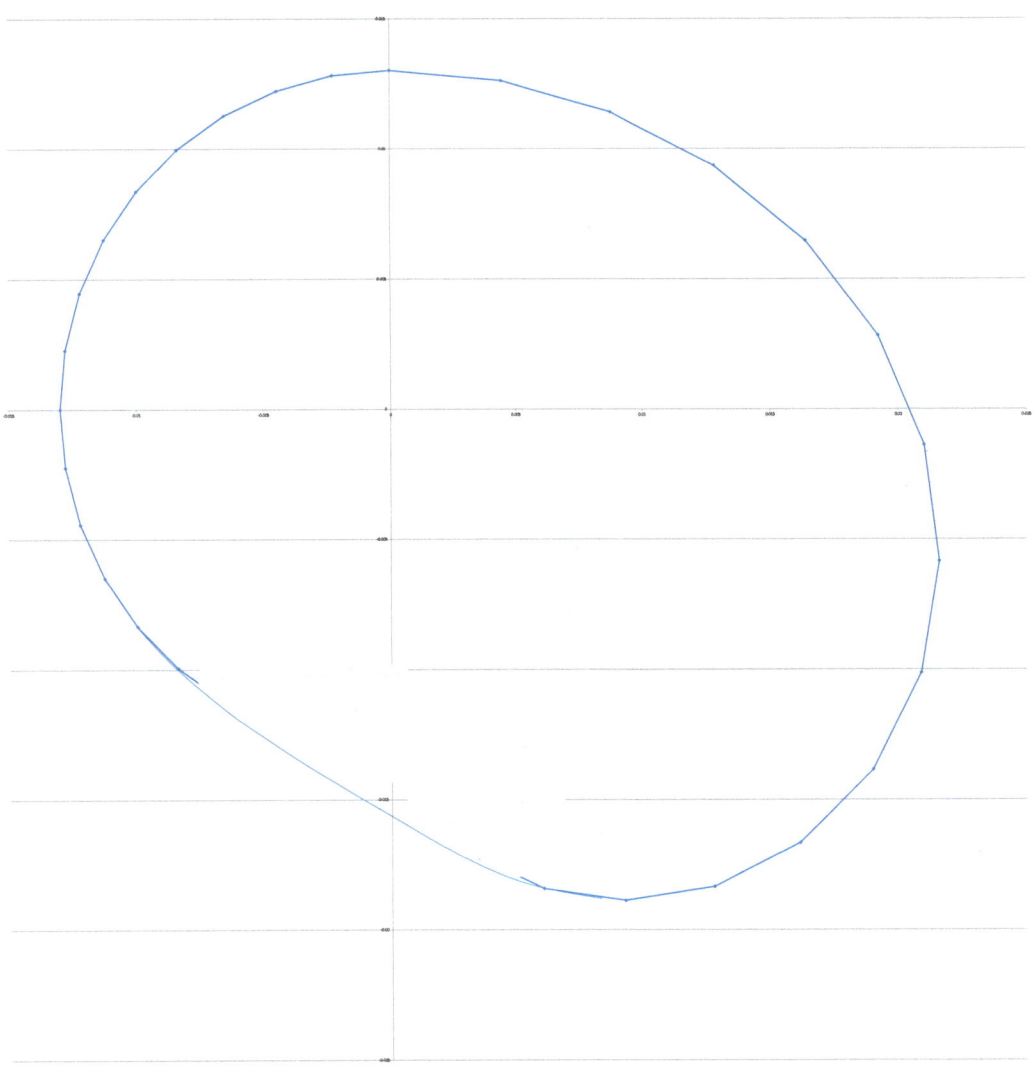

Fig. 5 *The closed profile of cam to the cam with plate translated follower*

4. The dynamics of the cam

The moments of inertia is determined with the relationships from the system 8.

$$
\begin{cases}
J_c = \dfrac{1}{2} M_c \cdot r^2 = \dfrac{1}{2} M_c \cdot \left(r_0^2 + 4r_0 \cdot s + 2s^2 \right); \quad J_T = m_T \cdot s'^2 \\[4mm]
J^* = \dfrac{1}{2} M_c \cdot \left(r_0^2 + 4r_0 \cdot s + 2s^2 \right) + m_T \cdot s'^2 = \dfrac{1}{2} M_c \cdot r_0^2 + J; \\[4mm]
\text{with} \quad J = 2M_c \cdot r_0 \cdot s + M_c \cdot s^2 + m_T \cdot s'^2 \\[6mm]
J_m^* = \dfrac{1}{2} M_c \cdot r_0^2 + \dfrac{1}{2} J_M; \quad \text{with} \quad J_{\min} = 0 \quad \text{when} \quad s = 0, \ s' = 0; \\[4mm]
J_{Max} \equiv J_M = Max\left(2M_c \cdot r_0 \cdot s + M_c \cdot s^2 + m_T \cdot s'^2 \right) \Rightarrow \\[2mm]
\Rightarrow J' = 2M_c \cdot r_0 \cdot s' + 2M_c \cdot s \cdot s' + 2m_T \cdot s' \cdot s'' \\[4mm]
J' = 0 \Rightarrow
\begin{cases}
s' = 0 \quad s = h \Rightarrow J_M = 2M_c \cdot r_0 \cdot h + M_c \cdot h^2 \quad I \\[4mm]
M_c \cdot r_0 + M_c \cdot s + m_T \cdot s'' = 0 \ \Rightarrow \cos\varphi > 1 \Rightarrow \varphi \in \phi \quad II
\end{cases}
\Rightarrow \\[4mm]
\Rightarrow J_m^* = \dfrac{1}{2} M_c \cdot r_0^2 + \dfrac{1}{2} 2M_c \cdot r_0 \cdot h + \dfrac{1}{2} M_c \cdot h^2 \Rightarrow \\[4mm]
\Rightarrow J_m^* = \dfrac{1}{2} M_c \cdot r_0^2 + M_c \cdot r_0 \cdot h + \dfrac{1}{2} M_c \cdot h^2 = \dfrac{1}{2} M_C \cdot \left(r_0 + h \right)^2 \\[6mm]
J_m^* = \dfrac{1}{2} M_C \cdot \left(r_0 + h \right)^2
\end{cases}
\tag{8}
$$

The angular velocity ω, and the angular acceleration, are determined with the presented relationships (system 9).

$$
\begin{cases}
\omega^2 = \dfrac{J_m^*}{J^*} \cdot \omega_m^2; \quad \omega = \sqrt{\dfrac{J_m^*}{J^*}} \cdot \omega_m \\[5mm]
\varepsilon = -\dfrac{1}{2} \cdot \omega^2 \cdot \dfrac{J^{*'}}{J^*} \\[6mm]
J_m^* = \dfrac{1}{2} M_C \cdot \left(r_0 + h \right)^2 \\[4mm]
J^* = \dfrac{1}{2} M_c \cdot \left(r_0^2 + 4r_0 \cdot s + 2s^2 \right) + m_T \cdot s'^2 \\[4mm]
J^{*'} \equiv J' = 2M_c \cdot r_0 \cdot s' + 2M_c \cdot s \cdot s' + 2m_T \cdot s' \cdot s''
\end{cases}
\tag{9}
$$

For a classic cam and tappet mechanism (without valve) dynamic movement tappet is expressed by equation (10), who was presented and derived in Chapter 2 (equation 48), and now by canceling valve mass, will customize and reaching form below (10).

$$x = s - \frac{(K+k)\cdot m_T \cdot \omega^2 \cdot s'^2 + (k^2 + 2k\cdot K)\cdot s^2 + 2k\cdot x_0 \cdot (K+k)\cdot s}{2\cdot (K+k)^2 \cdot \left(s + \dfrac{k\cdot x_0}{K+k}\right)} \tag{10}$$

Where x is the dynamic movement of the pusher, while s is its normal, kinematics movement. K is the spring constant of the system, and k is the spring constant of the tappet spring. It note, with x_0 the tappet spring preload, with m_T the mass of the tappet, with ω the angular rotation speed of the cam (or camshaft), where s' is the first derivative in function of φ of the tappet movement, s. Differentiating twice successively, the expression (10) in the angle φ, we obtain a reduced tappet speed (equation 11), and reduced tappet acceleration (12).

$$\begin{cases} N = (K+k)\cdot m_T \cdot \omega^2 \cdot s'^2 + (k^2 + 2k\cdot K)\cdot s^2 + 2k\cdot x_0 \cdot (K+k)\cdot s \\[2mm] M = \left[(K+k)m_T\omega^2 \cdot 2s's'' + (k^2 + 2kK)\cdot 2ss' + 2kx_0(K+k)\cdot s'\right]\cdot \\[2mm] \quad \cdot \left(s + \dfrac{kx_0}{K+k}\right) - N\cdot s' \\[4mm] x' = s' - \dfrac{M}{2\cdot (K+k)^2 \cdot \left(s + \dfrac{kx_0}{K+k}\right)^2} \end{cases} \tag{11}$$

$$\begin{cases} N = (K+k)\cdot m_T \cdot \omega^2 \cdot s'^2 + (k^2 + 2k\cdot K)\cdot s^2 + 2k\cdot x_0 \cdot (K+k)\cdot s \\[3mm] M = \left[(K+k)m_T\omega^2 \cdot 2s's'' + (k^2 + 2kK)\cdot 2ss' + 2kx_0(K+k)\cdot s'\right]\cdot \\[3mm] \quad \cdot \left(s + \dfrac{kx_0}{K+k}\right) - N\cdot s' \\[3mm] O = (K+k)\cdot m_T \cdot \omega^2 \cdot 2\cdot \left(s''^2 + s'\cdot s'''\right) + \\[2mm] \quad + (k^2 + 2\cdot k\cdot K)\cdot 2\cdot \left(s'^2 + s\cdot s''\right) + 2\cdot k\cdot x_0 \cdot (K+k)\cdot s'' \\[3mm] x'' = s'' - \dfrac{\left[O\cdot \left(s + \dfrac{kx_0}{K+k}\right) - N\cdot s''\right]\cdot \left(s + \dfrac{kx_0}{K+k}\right) - M\cdot 2\cdot s'}{2\cdot (K+k)^2 \cdot \left(s + \dfrac{kx_0}{K+k}\right)^3} \end{cases} \tag{12}$$

The real and dynamic, tappet acceleration can be determined directly using the relation (13).

$$\ddot{x} = x'' \cdot \omega^2 + x' \cdot \varepsilon \qquad (13)$$

For a law cos (14) we obtain the dynamic diagram from the Figure 6.

$$\begin{cases} s = \dfrac{h}{2} - \dfrac{h}{2} \cdot \cos\left(\pi \cdot \dfrac{\varphi}{\varphi_u} \right) \\[2mm] s' \equiv v_r = \dfrac{\pi \cdot h}{2 \cdot \varphi_u} \cdot \sin\left(\pi \cdot \dfrac{\varphi}{\varphi_u} \right) \\[2mm] s'' \equiv a_r = \dfrac{\pi^2 \cdot h}{2 \cdot \varphi_u^2} \cdot \cos\left(\pi \cdot \dfrac{\varphi}{\varphi_u} \right) \\[2mm] s''' \equiv \alpha_r = -\dfrac{\pi^3 \cdot h}{2 \cdot \varphi_u^3} \cdot \sin\left(\pi \cdot \dfrac{\varphi}{\varphi_u} \right) \end{cases} \qquad (14)$$

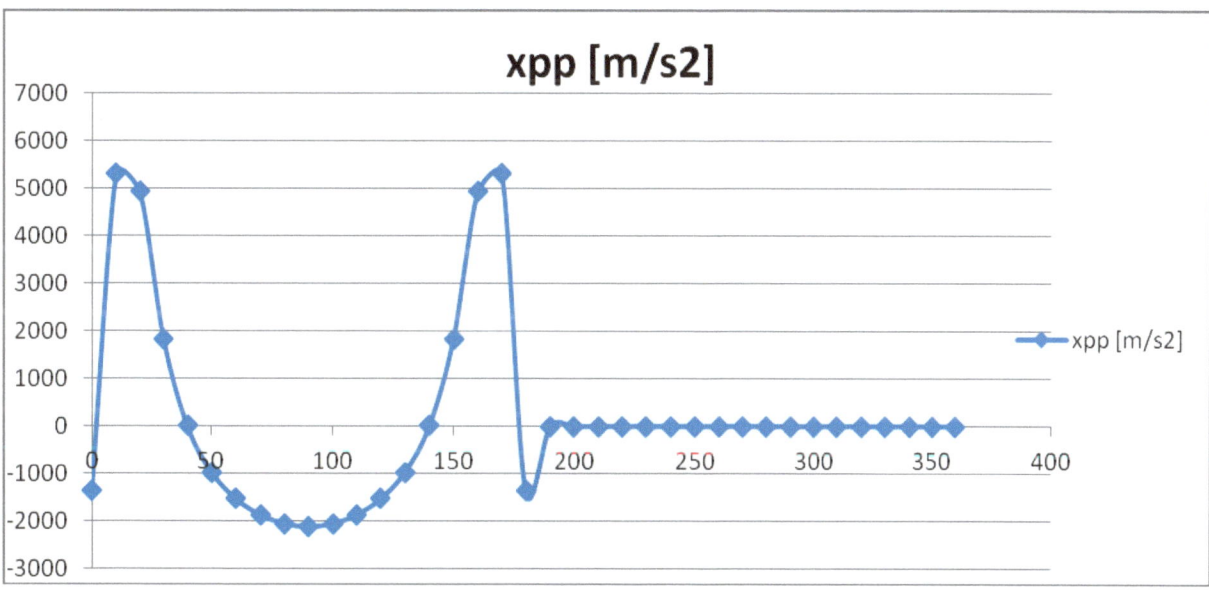

Fig. 6 *The dynamic diagram of the tappet acceleration from a cos profile of cam used to the cam with plate translated follower; r_0=13 [mm], M_c=200 [g], m_T=100 [g], φ_u=φ_c=π/2, h=6 [mm], n_m=10000 [rpm], x_0=90 [mm], k=40 [kN/m], K=5000 [kN/m].*

5. Conclusions

The presented dynamic system has the advantage to has a normal functionality. The synthesis was made using the natural geometro-kinematics parameters (of cam mechanism).

CHAPTER VII

DYNAMIC SYNTHESIS OF THE ROTARY CAM AND TRANSLATED TAPPET WITH ROLL

Abstract: This chapter presents an original methods to determine the dynamic parameters at the camshaft (the distribution mechanisms). We determine initially the mass moment of inertia (mechanical) of the mechanism, reduced to the element of rotation, ie at cam (basically using kinetic energy conservation, the system 1). The rotary cam with translated follower with roll (Figure 1), is synthesized dynamic. We considered the law of motion of the tappet classic version already used the cosine law (both ascending and descending). The angular velocity is a function of the cam position (φ) but also its rotation speed (2). Where ω_m is the nominal angular velocity of cam and express at the distribution mechanisms based on the motor shaft speed (3). We start the simulation with a classical law of motion, namely the cosine law. To climb cosine law system is expressed by relations (4). With the relation (5) is expressed the first derivative of the reduced mechanical moment of inertia. It is necessary to determine the angular acceleration (6). Relations (2) and (6) a general nature and is basically two original equations of motion crucial for mechanical mechanisms. For a rotary cam and translated tappet with roll mechanism (without valve), dynamic movement tappet is expressed by equation (7), who was presented and derived in Chapter 2 (equation 48), and now by canceling valve mass, will customize and reaching form below (7). Where x is the dynamic movement of the pusher, while s is its normal, kinematics movement. K is the spring constant of the system, and k is the spring constant of the tappet spring. It note, with x_0 the tappet spring preload, with m_T the mass of the tappet, with ω the angular rotation speed of the cam (or camshaft), where s' is the first derivative in function of φ of the tappet movement, s. Differentiating twice successively, the expression (7) in the angle φ, we obtain a reduced tappet speed (equation 8), and reduced tappet acceleration (9). Further the acceleration of the tappet can be determined directly real (dynamic) using the relation (10). For a good work one proposes to make a new geometro-kinematics synthesis of the cam profile, using some new relationships (16).

Keywords: cam, cams, cam mechanisms, distribution mechanisms, camshaft, tappet.

The rotary cam with translated follower with roll (Figure 1), is synthesized dynamic by the next relationships.

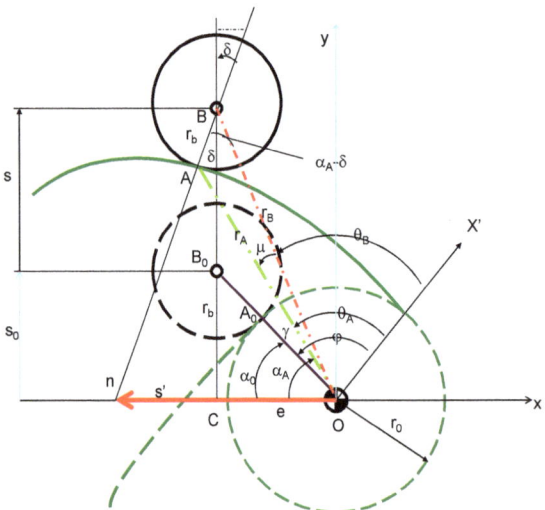

Fig. 1 *The rotary cam with translated follower with roll*

First, one determines the mass moment of inertia (mechanical) of the mechanism, reduced to the element of rotation, ie cam (basically using kinetic energy conservation, **system 1**).

$$\begin{cases}
J_{cama} = \dfrac{1}{2} \cdot M_c \cdot R^2 \\[4pt]
R^2 \equiv r_A^2 = x_A^2 + y_A^2 = e^2 + r_b^2 \cdot \sin^2 \delta + 2 \cdot e \cdot r_b \cdot \sin \delta + \\[4pt]
+ \left(s_0 + s\right)^2 + r_b^2 \cdot \cos^2 \delta - 2 \cdot r_b \cdot \left(s_0 + s\right) \cdot \cos \delta \\[4pt]
r_A^2 = e^2 + r_b^2 + \left(s_0 + s\right)^2 + 2 \cdot r_b \cdot \left[e \cdot \sin \delta - \left(s_0 + s\right) \cdot \cos \delta\right] \\[4pt]
r_A^2 = e^2 + r_b^2 + \left(s_0 + s\right)^2 + 2 \cdot r_b \cdot e \cdot \dfrac{s' - e}{\sqrt{\left(s_0 + s\right)^2 + \left(s' - e\right)^2}} - \\[4pt]
- 2 \cdot r_b \cdot \left(s_0 + s\right) \cdot \dfrac{\left(s_0 + s\right)}{\sqrt{\left(s_0 + s\right)^2 + \left(s' - e\right)^2}} \\[4pt]
r_A^2 = e^2 + r_b^2 + \left(s_0 + s\right)^2 - \dfrac{2 \cdot r_b \cdot \left(s_0 + s\right)^2}{\sqrt{\left(s_0 + s\right)^2 + \left(s' - e\right)^2}} + \\[4pt]
+ \dfrac{2 \cdot r_b \cdot e \cdot \left(s' - e\right)}{\sqrt{\left(s_0 + s\right)^2 + \left(s' - e\right)^2}} \\[4pt]
J_m^* = \dfrac{1}{2} \cdot M_c \cdot \left(r_0^2 + r_b^2 + r_0 \cdot r_b\right) + \dfrac{1}{4} \cdot M_c \cdot s_0 \cdot h + \dfrac{1}{16} \cdot M_c \cdot h^2 + \\[4pt]
+ \dfrac{1}{2} \cdot M_c \cdot r_b \cdot \dfrac{e \cdot \dfrac{\pi \cdot h}{2 \cdot \varphi_0} - e^2 - \left(s_0 + \dfrac{h}{2}\right)^2}{\sqrt{\left(s_0 + \dfrac{h}{2}\right)^2 + \left(\dfrac{\pi \cdot h}{2 \cdot \varphi_0} - e\right)^2}} + \dfrac{m_T \cdot \pi^2 \cdot h^2}{8 \cdot \varphi_0^2} \\[4pt]
J^* = \dfrac{1}{2} \cdot M_c \cdot \left(2 \cdot r_b^2 + r_0^2 + 2 \cdot r_0 \cdot r_b\right) + M_c \cdot s_0 \cdot s + \dfrac{1}{2} \cdot M_c \cdot s^2 + \\[4pt]
+ M_c \cdot r_b \cdot \dfrac{e \cdot s' - e^2 - \left(s_0 + s\right)^2}{\sqrt{\left(s_0 + s\right)^2 + \left(s' - e\right)^2}} + m_T \cdot s'^2
\end{cases} \qquad (1)$$

We considered the law of motion of the tappet classic version already used the cosine law (both ascending and descending).

The angular velocity is a function of the cam position (φ) but also its rotation speed (2). Where ω_m is the nominal angular velocity of cam and express at the distribution mechanisms based on the motor shaft speed (3).

$$\omega^2 = \frac{J_m^*}{J^*} \cdot \omega_m^2 \tag{2}$$

$$\omega_m = 2 \cdot \pi \cdot v_c = 2 \cdot \pi \cdot \frac{n_c}{60} = \frac{2 \cdot \pi}{60} \cdot \frac{n_{motor}}{2} = \frac{\pi \cdot n}{60} \tag{3}$$

We start the simulation with a classical law of motion, namely the cosine law. To climb cosine law system is expressed by relations (4).

$$\begin{cases} s = \dfrac{h}{2} - \dfrac{h}{2} \cdot \cos\left(\pi \cdot \dfrac{\varphi}{\varphi_u}\right) \\[2mm] s' \equiv v_r = \dfrac{\pi \cdot h}{2 \cdot \varphi_u} \cdot \sin\left(\pi \cdot \dfrac{\varphi}{\varphi_u}\right) \\[2mm] s'' \equiv a_r = \dfrac{\pi^2 \cdot h}{2 \cdot \varphi_u^2} \cdot \cos\left(\pi \cdot \dfrac{\varphi}{\varphi_u}\right) \\[2mm] s''' \equiv \alpha_r = -\dfrac{\pi^3 \cdot h}{2 \cdot \varphi_u^3} \cdot \sin\left(\pi \cdot \dfrac{\varphi}{\varphi_u}\right) \end{cases} \tag{4}$$

Where φ takes values from 0 to φ_u.

J_{max} occurs for $\varphi = \varphi_u/2$.

With the relation (5) is expressed the first derivative of the reduced mechanical moment of inertia. It is necessary to determine the angular acceleration (6).

$$\begin{aligned} J^{*'} = {}& M_c \cdot s_0 \cdot s' + M_c \cdot s \cdot s' + 2 \cdot m_T \cdot s' \cdot s'' + \\ & + M_c \cdot r_b \cdot \frac{\left[e \cdot s'' - 2 \cdot (s_0 + s) \cdot s'\right] \cdot \left[(s_0 + s)^2 + (s' - e)^2\right]}{\left[(s_0 + s)^2 + (s' - e)^2\right]^{3/2}} - \\ & - M_c \cdot r_b \cdot \frac{\left[e \cdot s' - e^2 - (s_0 + s)^2\right] \cdot \left[(s_0 + s) \cdot s' + (s' - e) \cdot s''\right]}{\left[(s_0 + s)^2 + (s' - e)^2\right]^{3/2}} \end{aligned} \tag{5}$$

Differentiating the formula (2), against time, is obtained the angular acceleration expression (6).

$$\varepsilon = -\frac{\omega^2}{2} \cdot \frac{J^{*'}}{J^*} \qquad (6)$$

Relations (2) and (6) a general nature and is basically two original equations of motion crucial for mechanical mechanisms.

For a rotary cam and translated tappet with roll mechanism (without valve), dynamic movement tappet is expressed by equation (7), who was presented and derived in Chapter 2 (equation 48), and now by canceling valve mass, will customize and reaching form below (7).

$$x = s - \frac{(K+k) \cdot m_T \cdot \omega^2 \cdot s'^2 + (k^2 + 2k \cdot K) \cdot s^2 + 2k \cdot x_0 \cdot (K+k) \cdot s}{2 \cdot (K+k)^2 \cdot \left(s + \dfrac{k \cdot x_0}{K+k} \right)} \qquad (7)$$

Where x is the dynamic movement of the pusher, while s is its normal, kinematics movement. K is the spring constant of the system, and k is the spring constant of the tappet spring.

It note, with x_0 the tappet spring preload, with m_T the mass of the tappet, with ω the angular rotation speed of the cam (or camshaft), where s' is the first derivative in function of φ of the tappet movement, s. Differentiating twice successively, the expression (7) in the angle φ, we obtain a reduced tappet speed (equation 8), and reduced tappet acceleration (9).

$$\left\{ \begin{array}{l} N = (K+k) \cdot m_T \cdot \omega^2 \cdot s'^2 + (k^2 + 2k \cdot K) \cdot s^2 + 2k \cdot x_0 \cdot (K+k) \cdot s \\[4mm] M = \left[(K+k)m_T\omega^2 \cdot 2s's'' + (k^2 + 2kK) \cdot 2ss' + 2kx_0(K+k) \cdot s' \right] \cdot \\[2mm] \quad \cdot \left(s + \dfrac{kx_0}{K+k} \right) - N \cdot s' \\[8mm] x' = s' - \dfrac{M}{2 \cdot (K+k)^2 \cdot \left(s + \dfrac{kx_0}{K+k} \right)^2} \end{array} \right. \qquad (8)$$

$$
\begin{cases}
N = (K+k) \cdot m_T \cdot \omega^2 \cdot s'^2 + (k^2 + 2k \cdot K) \cdot s^2 + 2k \cdot x_0 \cdot (K+k) \cdot s \\[3mm]
M = \left[(K+k) m_T \omega^2 \cdot 2s's'' + (k^2 + 2kK) \cdot 2ss' + 2kx_0 (K+k) \cdot s' \right] \cdot \\[2mm]
\quad \cdot \left(s + \dfrac{kx_0}{K+k} \right) - N \cdot s' \\[5mm]
O = (K+k) \cdot m_T \cdot \omega^2 \cdot 2 \cdot \left(s''^2 + s' \cdot s''' \right) + \\[2mm]
\quad + (k^2 + 2 \cdot k \cdot K) \cdot 2 \cdot \left(s'^2 + s \cdot s'' \right) + 2 \cdot k \cdot x_0 \cdot (K+k) \cdot s'' \\[5mm]
x'' = s'' - \dfrac{\left[O \cdot \left(s + \dfrac{kx_0}{K+k} \right) - N \cdot s'' \right] \cdot \left(s + \dfrac{kx_0}{K+k} \right) - M \cdot 2 \cdot s'}{2 \cdot (K+k)^2 \cdot \left(s + \dfrac{kx_0}{K+k} \right)^3}
\end{cases}
\tag{9}
$$

Further the acceleration of the tappet can be determined directly real (dynamic) using the relation (10).

$$
\ddot{x} = x'' \cdot \omega^2 + x' \cdot \varepsilon
\tag{10}
$$

Dynamic synthesis

Give the following parameters:

$r_0 = 0.013$ [m]; $r_b = 0.005$ [m]; $h = 0.008$ [m]; $e = 0.01$ [m]; $x_0 = 0.03$ [m]; $\varphi_u = \pi/2$; $\varphi_c = \pi/2$; $K = 5000000$ [N/m]; $k = 20000$ [N/m]; $m_T = 0.1$ [kg]; $M_c = 0.2$ [kg]; $n_{motor} = 5500$ [rot/min].

To sum up dynamically based on a computer program, you can vary the input data until the corresponding acceleration is obtained (see Figure 2). It then summarizes the corresponding cam profile (Figure 3) using the relations (11).

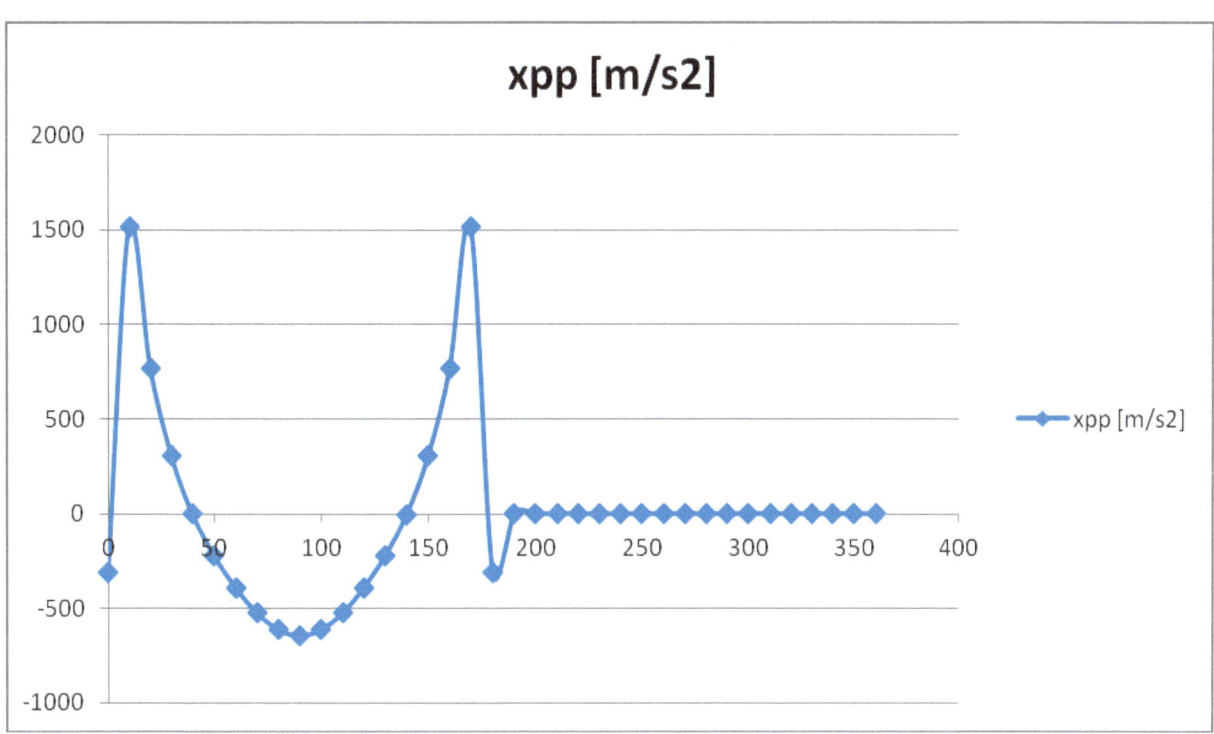

Fig. 2 *Dynamic diagram to the rotary cam with translated follower with roll*

$$
\begin{cases}
\begin{cases}
x_T = -e - r_b \cdot \sin \delta \\[2mm]
y_T = (s_0 + s) - r_b \cdot \cos \delta
\end{cases} \\[10mm]
\begin{cases}
x_C = x_T \cdot \cos \varphi - y_T \cdot \sin \varphi \\[2mm]
y_C = x_T \cdot \sin \varphi + y_T \cdot \cos \varphi
\end{cases} \\[10mm]
\begin{cases}
x_C = (-e - r_b \cdot \sin \delta) \cdot \cos \varphi - [(s_0 + s) - r_b \cdot \cos \delta] \cdot \sin \varphi \\[2mm]
y_C = (-e - r_b \cdot \sin \delta) \cdot \sin \varphi + [(s_0 + s) - r_b \cdot \cos \delta] \cdot \cos \varphi
\end{cases}
\end{cases} \tag{11}
$$

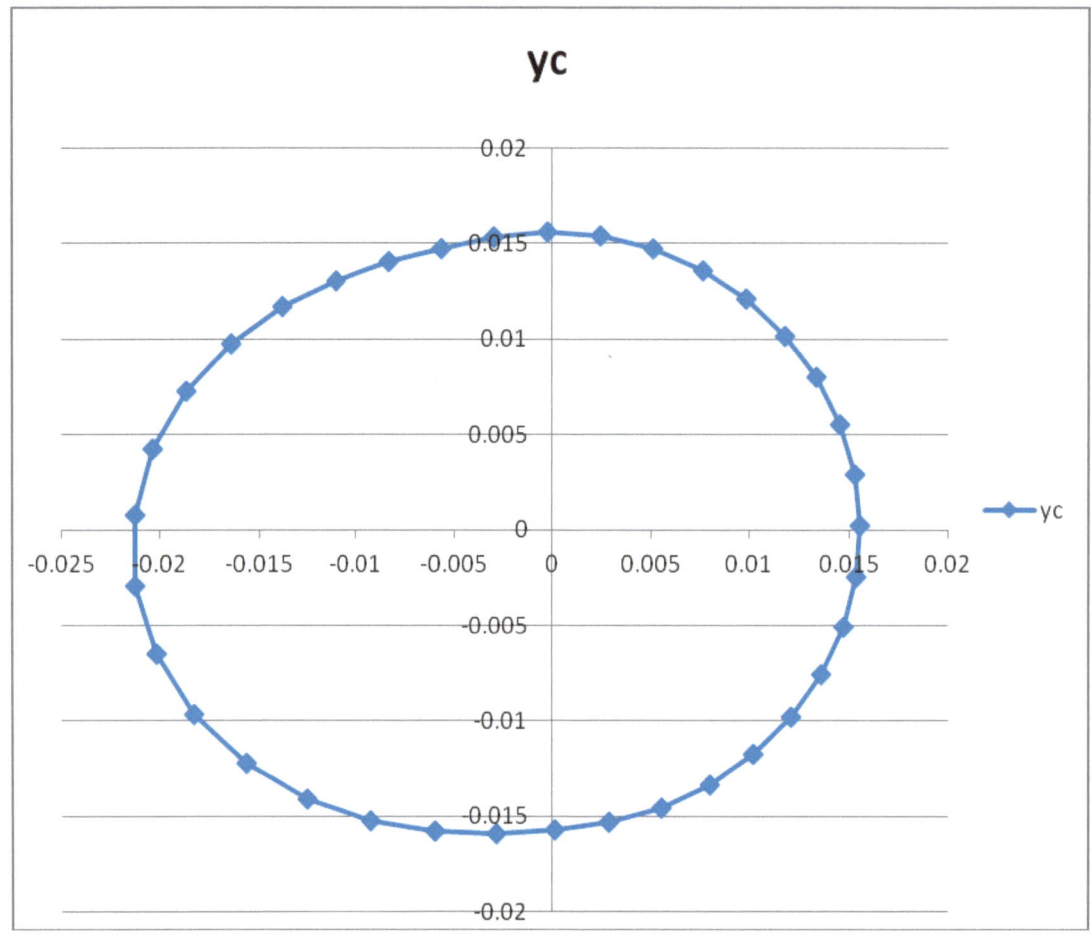

Fig. 3 *The cam profile to the rotary cam with translated follower with roll*

r_b=0.003 [m]; e=0.003 [m]; h=0.006 [m]; r_0=0.013 [m]; φ_0=π/2 [rad];

The geometry of the rotary cam and the translated follower with roll

Now, we shall see the geometry of a rotary cam with translated follower with roll (Figure 4). The cam rotation sense is positive (trigonometric).

We can make the geometrical synthesis of the cam profile with the help of the cinematics of the mechanism. One uses as well the reduced speed, s'.

OA=r=r_A; r^2=r_A^2

It establishes a system fixed Cartesian, xOy = $x_f Oy_f$, and a mobil Cartesian system, xOy = $x_m Oy_m$ fixed with the cam.

From the lower position 0, the tappet, pushed by cam, uplifts to a general position, when the cam rotates with the φ angle. The contact point A, go from A_i^0 to A^0 (on the cam), and to A (on the tappet). The position angle of the point A from the tappet is θ_f, and from the cam is θ_m. We can determine the coordinates of the point A from the tappet (12), and from the cam (13).

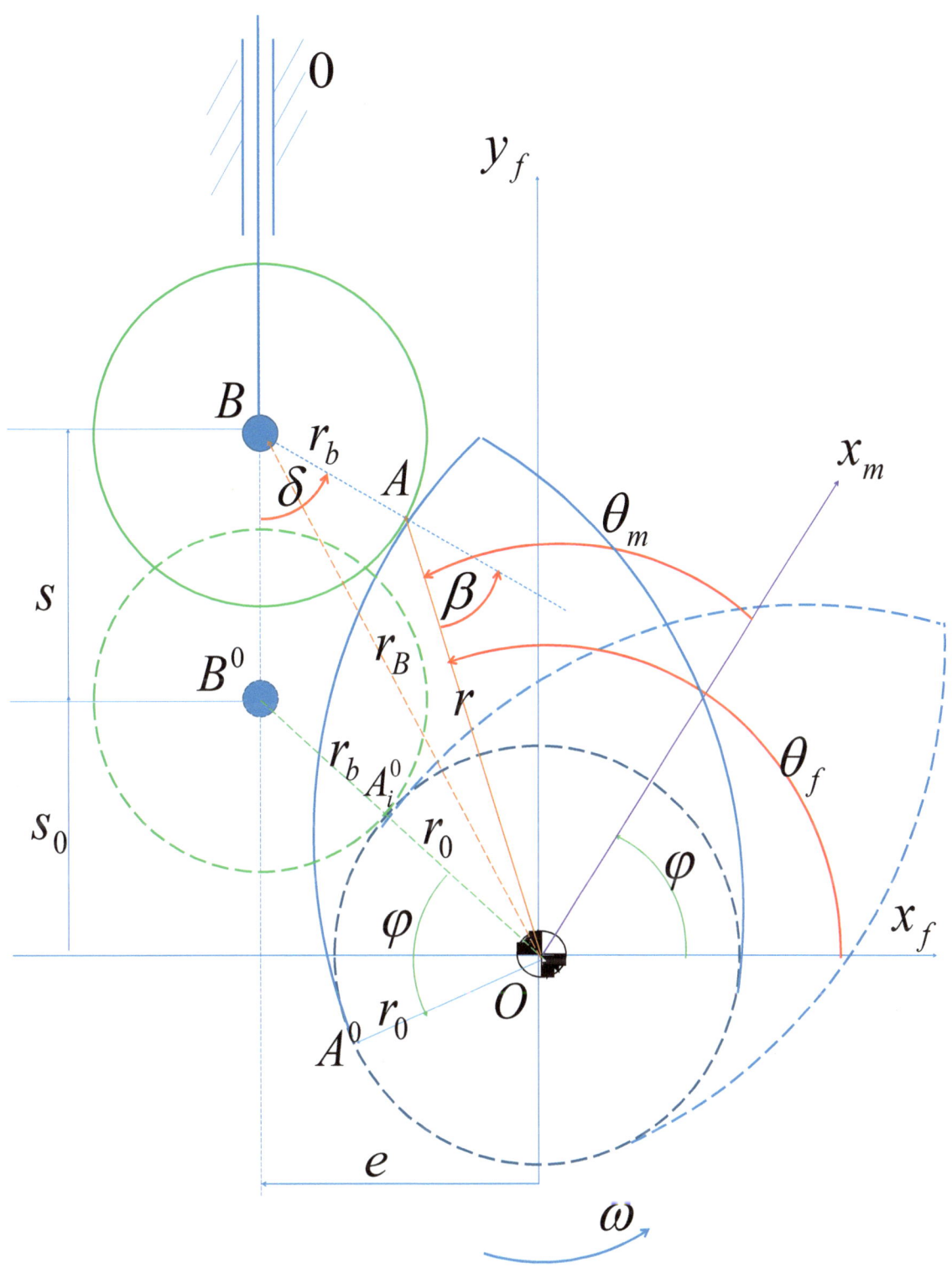

Fig. 4 *The geometry of the rotary cam with translated follower with roll*

$$\begin{cases} x_T \equiv x_A^f = -e + r_b \cdot \sin \delta = r_A \cdot \cos \theta_f = r \cdot \cos \theta_f \\ y_T \equiv y_A^f = s_0 + s - r_b \cdot \cos \delta = r_A \cdot \sin \theta_f = r \cdot \sin \theta_f \end{cases} \quad (12)$$

$$\begin{cases} x_c \equiv x_A^m = r_A \cdot \cos \theta_m = r \cdot \cos(\theta_f - \varphi) = r \cos \theta_f \cos \varphi + r \sin \theta_f \sin \varphi = \\ \quad = x_T \cos \varphi + y_T \sin \varphi = (-e + r_b \cdot \sin \delta) \cdot \cos \varphi + (s_0 + s - r_b \cdot \cos \delta) \cdot \sin \varphi \\ \\ y_c \equiv y_A^m = r_A \cdot \sin \theta_m = r \cdot \sin(\theta_f - \varphi) = r \sin \theta_f \cos \varphi - r \sin \varphi \cos \theta_f = \\ \quad = y_T \cos \varphi - x_T \sin \varphi = (s_0 + s - r_b \cdot \cos \delta) \cdot \cos \varphi - (-e + r_b \cdot \sin \delta) \cdot \sin \varphi \end{cases} \quad (13)$$

One uses and the next relationships (where the pressure angle δ was obtained with the classic Antonescu P. method):

$$\begin{cases} s_0 = \sqrt{(r_0 + r_b)^2 - e^2} \\ \\ \cos \delta = \dfrac{s_0 + s}{\sqrt{(s_0 + s)^2 + (s' - e)^2}} \\ \\ \sin \delta = \dfrac{s' - e}{\sqrt{(s_0 + s)^2 + (s' - e)^2}} \\ \\ tg\delta = \dfrac{s' - e}{s_0 + s} \end{cases} \quad (14)$$

Determining the forces, the velocities and the efficiency (15)

The driving force F_m, perpendicular on r in A, is divided in two components: F_n, the normal force, and F_a, a force of slipping. Fn is divided, as well, in two components: F_T is the transmitted (the utile) force, and F_R is a radial force which bend the tappet (see 15, and the Figure 5).

$$\begin{cases} \begin{cases} F_n = F_m \cdot \cos \alpha \\ v_n = v_m \cdot \cos \alpha \end{cases} \begin{cases} F_T = F_n \cdot \cos \delta = F_m \cdot \cos \alpha \cdot \cos \delta \\ v_T = v_n \cdot \cos \delta = v_m \cdot \cos \alpha \cdot \cos \delta \end{cases} \\ \\ \mu_i = \dfrac{P_u}{P_c} = \dfrac{F_T \cdot v_T}{F_m \cdot v_m} = \dfrac{F_m \cdot \cos \alpha \cdot \cos \delta \cdot v_m \cdot \cos \alpha \cdot \cos \delta}{F_m \cdot v_m} = (\cos \alpha \cdot \cos \delta)^2 = \cos^2 \alpha \cdot \cos^2 \delta \\ \\ \beta = \pi - A; \quad \alpha = \dfrac{\pi}{2} - \beta = A - \dfrac{\pi}{2}; \quad \cos A = \dfrac{r_b^2 + r^2 - r_B^2}{2 \cdot r_b \cdot r}; \quad r_B = \sqrt{e^2 + (s_0 + s)^2}; \\ \\ r = \sqrt{x_{Af}^2 + y_{Af}^2} = \sqrt{(-e + r_b \cdot \sin \delta)^2 + (s_0 + s - r_b \cdot \cos \delta)^2} \Rightarrow \\ \\ \Rightarrow r \equiv r_A = \sqrt{e^2 + r_b^2 + (s_0 + s)^2 - 2 \cdot r_b \cdot [e \cdot \sin \delta + (s_0 + s) \cdot \cos \delta]} \end{cases} \quad (15)$$

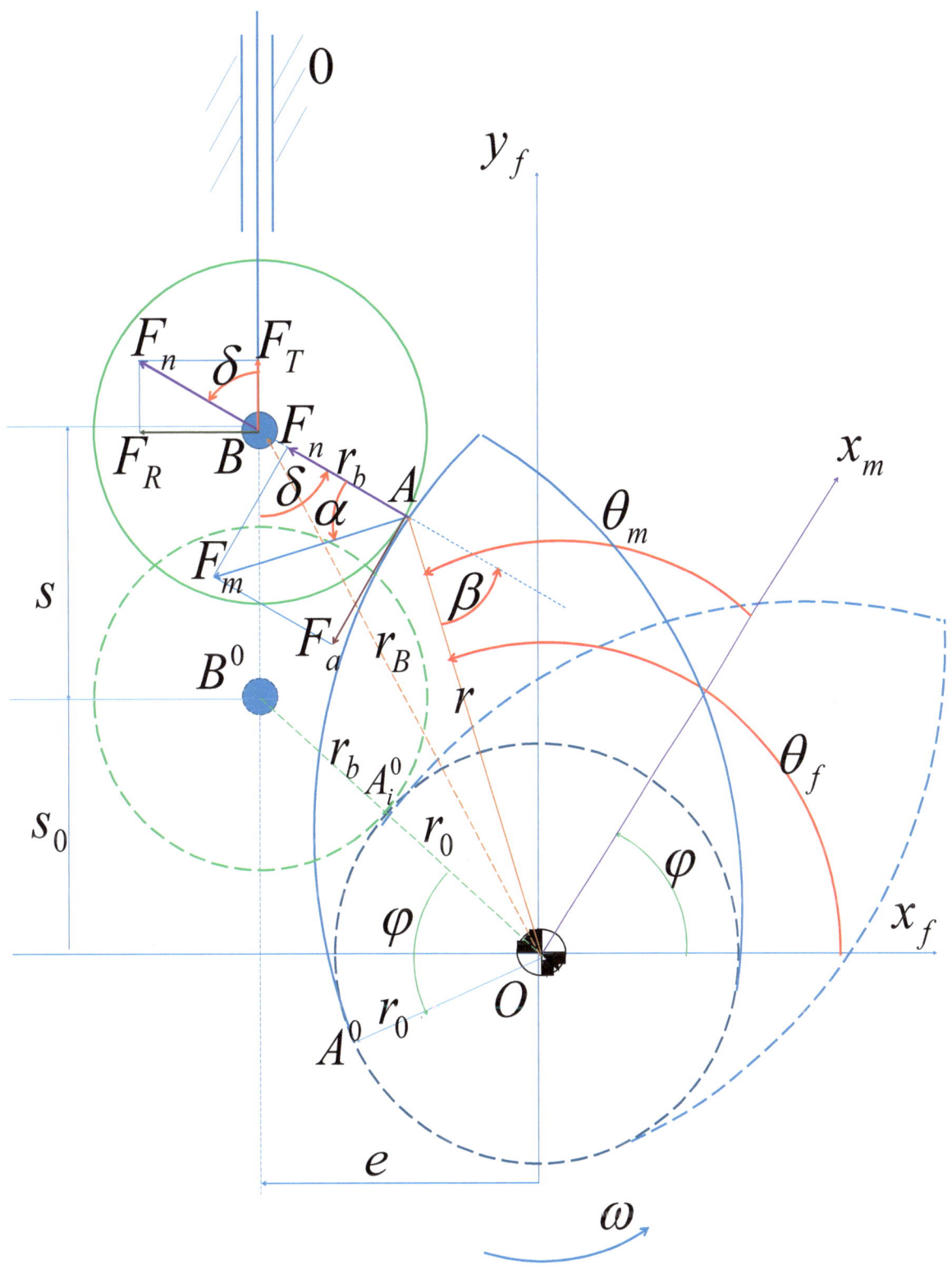

Fig. 5 *Forces and velocities of the rotary cam with translated follower with roll*

Geometro-kinematics synthesis

For a good work one proposes to make a new geometro-kinematics synthesis of the cam profile, using some new relationships (16).

$$\begin{cases} x_T \equiv x_A^f = -e + r_b \cdot \sin\delta = r_A \cdot \cos\theta_f = r \cdot \cos\theta_f \\ y_T \equiv y_A^f = s_0 + s - r_b \cdot \cos\delta = r_A \cdot \sin\theta_f = r \cdot \sin\theta_f \end{cases} \qquad (12)$$

$$\begin{cases} x_c \equiv x_A^m = r_A \cdot \cos\theta_m = r \cdot \cos(\theta_f - \varphi) = r\cos\theta_f \cos\varphi + r\sin\theta_f \sin\varphi = \\ = x_T \cos\varphi + y_T \sin\varphi = (-e + r_b \cdot \sin\delta) \cdot \cos\varphi + (s_0 + s - r_b \cdot \cos\delta) \cdot \sin\varphi \\ \\ y_c \equiv y_A^m = r_A \cdot \sin\theta_m = r \cdot \sin(\theta_f - \varphi) = r\sin\theta_f \cos\varphi - r\sin\varphi\cos\theta_f = \\ = y_T \cos\varphi - x_T \sin\varphi = (s_0 + s - r_b \cdot \cos\delta) \cdot \cos\varphi - (-e + r_b \cdot \sin\delta) \cdot \sin\varphi \end{cases} \qquad (13)$$

One uses and the next relationships (where the pressure angle δ was obtained with the new Petrescu F. method):

$$\begin{cases} s_0 = \sqrt{(r_0 + r_b)^2 - e^2} \\ \\ \\ \cos\delta = \sqrt{\dfrac{(s_0 + s)^2 + (s_0 + s) \cdot \sqrt{(s_0 + s)^2 - 4 \cdot s'^2} - 4 \cdot e \cdot s' - 2 \cdot e \cdot s'}{2 \cdot [(s_0 + s)^2 + e^2]}} \\ \Rightarrow \delta = \arccos(\cos\delta) \Rightarrow \\ \\ \delta = \arccos\left(\sqrt{\dfrac{(s_0 + s)^2 + (s_0 + s) \cdot \sqrt{(s_0 + s)^2 - 4 \cdot s'^2} - 4 \cdot e \cdot s' - 2 \cdot e \cdot s'}{2 \cdot [(s_0 + s)^2 + e^2]}} \right) \\ \\ \\ \sin\delta = \sin\left(\arccos\left(\sqrt{\dfrac{(s_0 + s)^2 + (s_0 + s) \cdot \sqrt{(s_0 + s)^2 - 4 \cdot s'^2} - 4 \cdot e \cdot s' - 2 \cdot e \cdot s'}{2 \cdot [(s_0 + s)^2 + e^2]}} \right) \right) \end{cases} \qquad (16)$$

The new profile can be seen in the Figure 6.

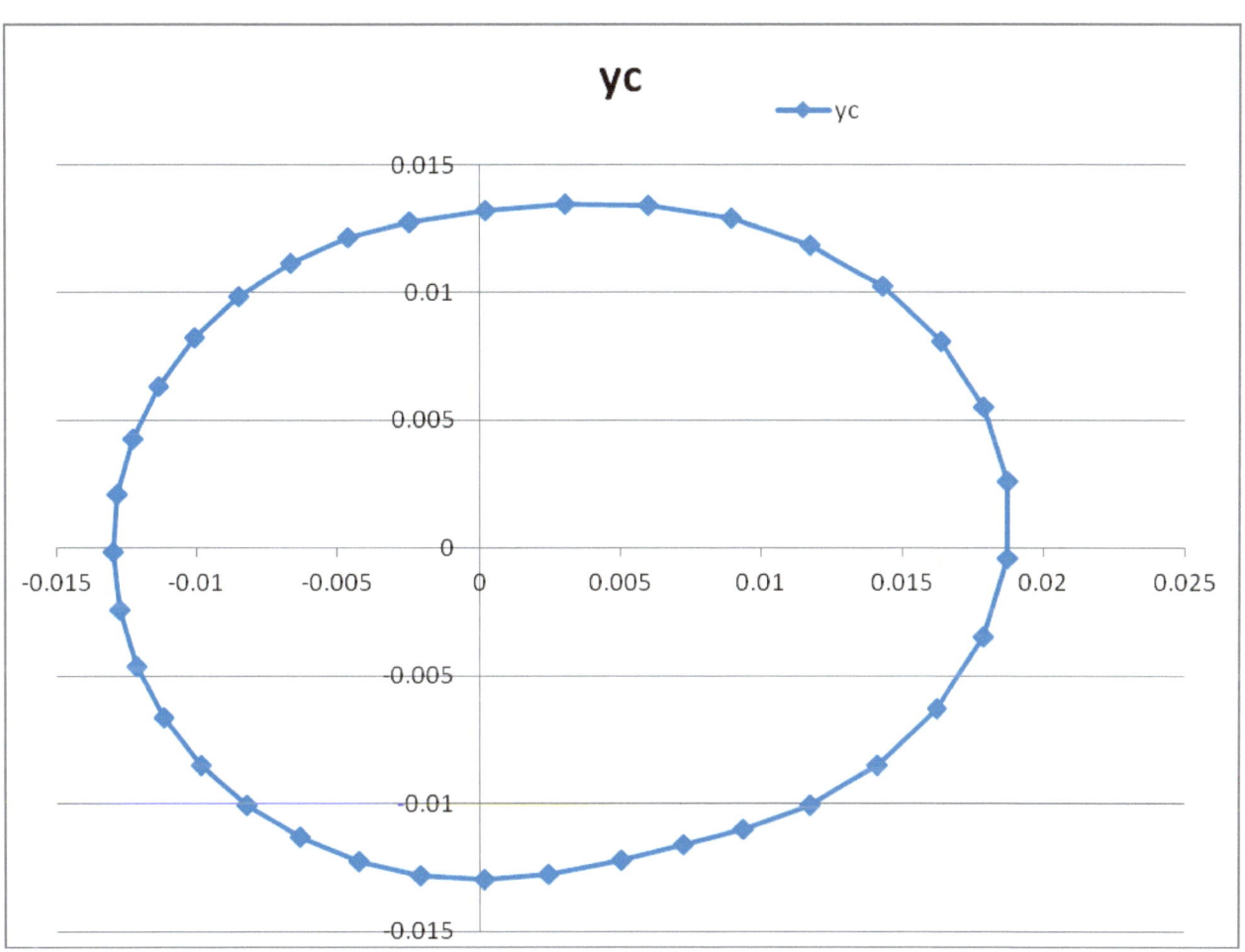

Fig. 6 *The new cam profile to the rotary cam with translated follower with roll*
r_b=0.003 [m]; e=0.003 [m]; h=0.006 [m]; r_0=0.013 [m]; φ_0=π/2 [rad];

Demonstration (explication)

$$\dot{s} = r_A \cdot \omega \cdot \cos\alpha \cdot \cos\delta \Rightarrow s' = r_A \cdot \cos\alpha \cdot \cos\delta$$

$$\alpha + \beta = \frac{\pi}{2}; \quad A + \beta = \pi; \quad \alpha = \frac{\pi}{2} - \beta = A - \frac{\pi}{2} \Rightarrow$$

$$\Rightarrow \cos\alpha = \cos\left(A - \frac{\pi}{2}\right) = \cos\left(\frac{\pi}{2} - A\right) = \sin A$$

$$\frac{s'}{r_A \cdot \cos\delta} = \sin A$$

$$\sin A = \frac{r_B}{r_A} \cdot \sin(\delta - B)$$

$$\Rightarrow \frac{s'}{r_A \cdot \cos\delta} = \frac{r_B}{r_A} \cdot \sin(\delta - B) \Rightarrow \frac{s'}{r_B \cdot \cos\delta} = \sin(\delta - B)$$

$$\sin(\delta - B) = \sin\delta\cos B - \sin B\cos\delta = \frac{s_0 + s}{r_B}\sin\delta - \frac{e}{r_B}\cos\delta$$

$$\Rightarrow \frac{s'}{r_B \cdot \cos\delta} = \frac{(s_0 + s)\cdot\sin\delta - e\cdot\cos\delta}{r_B} \Rightarrow s' = (s_0 + s)\cdot\sin\delta\cdot\cos\delta - e\cdot\cos^2\delta \Rightarrow$$

$$\Rightarrow (s_0 + s)\cdot\cos\delta\cdot\sqrt{1 - \cos^2\delta} = s' + e\cdot\cos^2\delta \Rightarrow$$

$$\Rightarrow (s_0 + s)^2\cdot\cos^2\delta - (s_0 + s)^2\cdot\cos^4\delta = s'^2 + e^2\cdot\cos^4\delta + 2\cdot e\cdot s'\cdot\cos^2\delta \Rightarrow$$

$$\Rightarrow \left[(s_0 + s)^2 + e^2\right]\cdot\cos^4\delta - \left[(s_0 + s)^2 - 2\cdot e\cdot s'\right]\cdot\cos^2\delta + s'^2 = 0 \Rightarrow$$

$$\Rightarrow \cos^2\delta = \frac{(s_0 + s)^2 - 2es' \pm \sqrt{\left[(s_0 + s)^2 - 2\cdot e\cdot s'\right]^2 - 4s'^2\left[(s_0 + s)^2 + e^2\right]}}{2\cdot\left[(s_0 + s)^2 + e^2\right]} \Rightarrow \tag{17}$$

$$\Rightarrow \cos\delta = \sqrt{\frac{(s_0 + s)^2 + (s_0 + s)\cdot\sqrt{(s_0 + s)^2 - 4\cdot s'^2} - 4\cdot e\cdot s' - 2\cdot e\cdot s'}{2\cdot\left[(s_0 + s)^2 + e^2\right]}}$$

Bibliography

1.-A1. ANTONESCU P., *Mecanisme - Calculul structural si cinematic*. I.P.B., Bucuresti, 1979.

2.-A2. ANTONESCU P., *Cinetostatica si dinamica mecanismelor*. I.P.B.,Bucuresti, 1980.

3.-A3. ANTONESCU P., *Sinteza mecanismelor*. I.P.B.,Bucuresti, 1983.

4.-A4. ANTONESCU P., COMANESCU A.,GRECU B., *Indrumar de proiect la mecanisme. Partea a I-a*, I.P.B., Bucuresti, 1987.

5.-A5. ALEXANDRU P., DUTA FL.,JULA A., *Mecanismele directiei autovehiculelor*. Editura tehnică, Bucuresti, 1977.

6.-A6. ARTOBOLEVSKI I., *Teoria mehanizov*, Izd. Nauka, Moskva, 1965.

7.-A7. ANTONESCU P., DRANGA M., TEMPEA I., *Asigurarea preciziei cinematice a preselor de vulcanizat camere de aer*. In revista Constructia de masini, nr.8., Bucuresti, 1978.

8.-A8. ATANASIU M., *Mecanica*. Ed. Did. Ped., Bucuresti, 1973.

9.-A9. ATTILA H., DRAGULESCU D., *Probleme de mecanică - dinamică*. Editura Helicon, Timisoara, 1993.

10.-A10. ANTONESCU P., *Sinteza mecanismului cu camă rotativă si tachet translant*. In al V-lea Simpozion national de mecanisme si transmisii mecanice, Cluj-Napoca, 20-22 octombrie 1988.

11.-A11. ANTONESCU P., PETRESCU FL., *Metodă analitică de sinteză a mecanismului cu camă si tachet plat*. In al IV-lea Simpozion international de teoria si practica mecanismelor, Vol. III-1., Bucuresti, iulie 1985.

12.-A12. ANTONESCU P., OPREAN M., PETRESCU FL., *Contributii la sinteza mecanismului cu camă oscilantă si tachet plat oscilant*. In al IV-lea Simpozion international de teoria si practica mecanismelor, Vol. III-1., Bucuresti, iulie 1985.

13.-A13. ANTONESCU P., OPREAN M., PETRESCU FL., *La projection de la came oscillante chez les mechanismes a distribution variable*. In a V-a Conferintă de motoare, automobile, tractoare si masini agricole, Vol. I-motoare si automobile, Brasov, noiembrie 1985.

14.-A14. ANTONESCU P., OPREAN M., PETRESCU FL., *Proiectarea profilului Kurz al camei rotative ce actionează tachetul plat oscilant cu dezaxare*. In al III-lea Siopozion national de proiectare asistată de calculator în domeniul mecanismelor si organelor de masini-PRASIC'86, Brasov, decembrie 1986.

15.-A15. ANTONESCU P., OPREAN M., PETRESCU FL., *Analiza dinamică a mecanismelor de distributie cu came*. In al VII-lea Simpozion national de roboti industriali si mecanisme spatiale, Vol. 3., Bucuresti, octombrie 1987.

16.-A16. ANTONESCU P., OPREAN M., PETRESCU FL., *Sinteza analitică a profilului Kurz, la cama cu tachet plat rotativ*. In revista Constructia de masini, nr. 2., Bucuresti, 1988.

17.-A17. ANTONESCU P., PETRESCU FL., *Contributii la analiza cinetoelastodinamică a mecanismelor de distributie*. In SYROM'89, Bucuresti, iulie 1989.

18.-A18. ANTONESCU P., PETRESCU FL., ANTONESCU O., *Contributii la sinteza mecanismului cu camă rotativă si tachet balansier cu vârf*. In PRASIC'94, Brasov, decembrie 1994.

19.-A19. AUTORENKOLLEKTIV (J. VOLMER COORDONATOR), *Getriebetechnik-VEB*, Verlag technik, pp. 345-390, Berlin, 1968.

20.-A20. ANGELAS J., LOPEZ-CAJUN C., *Optimal synthesis of cam mechanisms with oscillating flat-face followers*. Mechanism and Machine Theory 23,(1988), Nr. 1., pp. 1-6., 1988.

21.-A21. ARAMA C., SERBANESCU A., *Economia de combustibil la automobile*. Editura tehnică, Bucuresti, 1974.

22.-A22. ALLAIS D.C., *Cycloidal vs modified trapezoid cams*. Machine Design 35(3), 31 Jan. 1963, pp. 92-96.

23.-A23. ANDERSON D.G., *Cam dynamics*. Prod. Engineering, 24(10), 1953, pp. 170-176.

24.-A24. ASTROP A.W., *Automatic high-speed inspection of variable pitch cams for zoom lenses*. Machinery (London), 1967, 110(2849), pp. 1360-1364.

25.-A25. AOYAGI Y., s.a., Hino Motors, Ltd. Japan, *Swirl Formation Process in Four Valve Diesel Engines*. (945011), In XXV FISITA Congres, 17-21 October 1994, Beijing, pp. 99-105.

26.-A26. ANTONESCU P., sa., *Contributions to the synthesis of the oscillating cam profile in the variable distribution mechanisms,* Eighth World Congress on TMM, Praga, vol. 5, 1991.

27.-A27. ANTONESCU P., PETRESCU FL., ANTONESCU D., *Geometrical synthesis of the rotary cam and balance tappet mechanism*. SYROM'97, Vol. 3, pp. 23, Bucuresti, august 1997.

28.-A28. ANTONESCU P., *Mecanisme și Manipulatoare, aplicaţii-teme de proiect*, Printech, Buc., 2000.

29.-A29. ANTONESCU, P., PETRESCU, F., ANTONESCU, O. *Contributions to the Synthesis of The Rotary Disc-Cam Profile,* In VIII-th International Conference on the Theory of Machines and Mechanisms, Liberec, Czech Republic, pp. 51-56, 2000.

30.-A30. ANTONESCU, P., PETRESCU, F., ANTONESCU, O., *Synthesis of the Rotary Cam Profile with Balance Follower,* In the 8-th Symposium on Mechanisms and Mechanical Transmissions, Timişoara, Vol. 1, pp. 39-44, 2000.

31.-A31. ANTONESCU, P., PETRESCU, F., ANTONESCU, O. *Contributions to the synthesis of mechanisms with rotary disc-cam.* In The Eigth IFToMM International Symposium on Theory of Machines and Mechanisms, SYROM'2001, Bucharest, ROMANIA, 2001, Vol. III, p. 31-36.

32.-A32. ANTONESCU P., OCNARESCU C., ANTONESCU O., *Mecanisme și Manipulatoare-îndrumar de laborator,* Ed. Printech, Bucuresti, 2002.

33.-A33. ANTONESCU P., *Sinteza unitară geometro-cinematică a profilului camei-disc rotative*, Rev. Mecanisme și Manipulatoare, I, 2, 2002.

34.-A34. ANTONESCU P., sa., *Geometric and Kinematic Synthesis of Mechanisms with Rotary Disc-Cam,* Proceedings of the 11th World Congress in Mechanism and Machine Science, Tianjin, 2003.

35.-A35. ANTONESCU P., *Mecanisme,* Printech, Bucuresti, 2003.

36.-A36. ANTONESCU P., *Mechanism and Machine Science*, Printech Press, Bucharest, Romania, 2005.

37.-A37. ANTONESCU P., ANTONESCU O., *Aplicaţii de mecanică tehnică, mecanisme și manipulatoare,* Printech, 2007.

38.-B1. BUZDUGAN GH., *Teoria vibratiilor si aplicatiile ei în constructia de masini*. Editura tehnică, Bucuresti, 1958.

39.-B2. BUZDUGAN GH., *Rezistenta materialelor*. Editura didactică si pedagogică, Bucuresti, 1964.

40.-B3. BOGDAN R., LARIONESCU D., CONONOVICI S., *Sinteza mecanismelor plane articulate*. Editura Academiei R.S.R., Bucuresti, 1977.

41.-B4. BOGDAN R., LARIONESCU D., *Analiza armonică complexă si mecano-electrică a mecanismelor plane*. Editura Academiei R.S.R., Bucuresti, 1968.

42.-B5. BALAN ST., *Probleme de mecanică*. Editura didactică si pedagogică, Bucuresti, 1977.

43.-B6. BUZDUGAN GH., FETCU L., RADES M., *Vibratii mecanice*. Editura didactică si pedagogică, Bucuresti, 1979.

44.-B7. BUZDUGAN GH., MIHAILESCU E., RADES M., *Măsurarea vibratiilor*. Editura Academiei R.S.R., Bucuresti, 1979.

45.-B8. BOBANCU S., *Consideratii cinetoelastice asupra variabilei "excentricitate" a mecanismelor plane cu camă având tachet oscilant plat*. In al IV-lea Simpozion international de teoria si practica mecanismelor, Vol. III-1., Bucuresti, iulie 1985.

46.-B9. BARSAN A., *Algoritm de sinteză asistată de calculator, a mecanismelor plane cu camă de rotatie si tachet plat*. In al VII-lea Simpozion national de roboti industriali si mecanisme spatiale. Vol. 3., Bucuresti, octombrie 1987.

47.-B10. BARSAN A., *Algoritm de sinteză asistată de calculator a mecanismelor cu camă cilindrică*. In al VII-lea Simpozion national de roboti industriali si mecanisme spatiale. Vol. 3., Bucuresti, octombrie 1987.

48.-B11. BOGDAN R., S.A., *Algoritm si program pentru analiza cinematică si dinamică a mecanismelor diferentiale complexe*. In al VII-lea Simpozion national de roboti industriali si mecanisme spatiale. Vol. 3., Bucuresti, octombrie 1987.

49.-B12. BUGAEVSKI E., *Contributii la studiul cinematic si dinamic al mecanismelor cu trenuri diferentiale*. Teză de doctorat, I.P.B., 1971.

50.-B13. BOIANGIU D., s.a., *Elemente elastice ale masinilor*. Editura tehnică, Bucuresti, 1967.

51.-B14. BUZDUGAN GH., *Izolarea antivibratorie a masinilor*. Editura Academiei R.S.R., Bucuresti, 1980.

52.-B15. BLOOM D., and RADCLIFFE C.W., *The effect of camshaft elasticity on the response of cam driven systems*, ASME paper 64-mech 41.

53.-B16. BARTON P., REESJONES J., *The dynamic effects of functional clearance and motor characteristics on the performance of a Geneva mechanism*. IFTOMM International Symp. on Linkages and Computer Design Methods, Bucharest, 1973.

54.-B17. BARABYI J.S., *Cams, dynamics and design*. Design News, 1969, 24, pp. 108.

55.-B18. BARKAN P., *Calculation of high-speed valve motion with flexible overhead linkage*. Trans. SAE, 1953, 61,pp. 687-700.

56.-B19. BEARD C.A., *Problems în valve gear design and instrumentation*. SAE Technical Progress Series, 1963, pp. 58-84.

57.-B20. BEARD C.A., *Cam mechanism design problems-an engine designer's view point*. In, Cams and cam mechanisms, Edited by J. REES JONES, MEP, London and Birmingham, Alabama, 1974, pp.49-53.

58.-B21. BARKAN P., s.a., *A spring-actuated, cam follower system; Design theory and experimental result*. Journal Engineering, Trans. ASME, 1965,(87 B), pp. 279-286.

59.-B22. BAUMGARTEN J.R., *Preload force necessary to prevent separation of follower from cam*. Trans. 7 th. Conf. on Mech., Purdue University, 1962.

60.-B23. BENEDICT C.A., s.a., *Dynamic responses of a mechanical system containing a coulomb friction force*. The 3 rd. Appl. Mech. Conf. Paper, Nr. 44., Oklahoma State University, 1973.

61.-B24. BAXTER M.L., *Qurvature-acceleration relation for plane cams*. Trans. ASME 70,1948, pp.483-489.

62.-B25. BISHOP J.L.H., *An analytical approach to automobile valve gear design*. Inst. of Mech. Engrs. Auto-Division Proc. 4, 1950-51, pp. 150-160.

63.-B26. BUHAYAR E.S., *Computerized cam design and plate cam manufacture*. Paper Nr. 66-MECH-2, ASME Mechanisms Conference, Lafayette, Ind., Oct. 1966.

64.-B27. BARBULESCU N., *Bazele fizice ale relativitătii Einsteiniene*. In E.S.E., Bucuresti, 1979.

65.-B28. BACKLUND O., s.a., *Volvo's MEP and PCP Engines: Combining Environmental Benefit with High Performance*. In Fifth Autotechnologies Conference Proceedings, SAE, (910010), pp. 238.

66.-C1. CHIRIACESCU S., *Proiectarea automată a camelor folosite la masina de ascutit pânze de fierăstrău*. In al IV-lea Simpozion international de teoria si practica mecanismelor, Vol. III-1., Bucuresti, iulie 1985.

67.-C2. CIONCA O., *Studiul mecanismelor camă-tachet ca sisteme oscilante autoexcitante*. In al IV-lea SYROM'85, Vol. III-1., Bucuresti, iulie 1985.

68.-C3. COMANESCU D., COMANESCU A.,S.A., *Sinteza profilelor zonelor de contact ale elementelor cinematice din mecanismele perforatoarelor de bandă*. In al IV-lea SYROM'85, Vol. III-1., Bucuresti, iulie 1985.

69.-C4. COMANESCU A., COMANESCU D., *Aplicarea sistemelor modulare de calcul cinetodinamic la instruirea si comanda mecanismelor multimobile*. In al VII-lea Simpozion national de roboti industriali si mecanisme spatiale, Vol. 3., Bucuresti, octombrie 1987.

70.-C5. CONSTANTINESCU G., *Teoria sonicitătii*. Ed. Academiei R.S.R., Bucuresti, 1985.

71.-C6. CRUDU M., *Contributii la studiul mecanismelor cu conexiuni dinamice*. Teză de doctorat, I.P.B., 1971.

72.-C7. CECCARELLI M., GARCIA-LOMAS J., *On the dynamics of two-link manipulators*. Al VI-lea SYROM, Vol. II.,Bucuresti, iunie 1993.

73.-C8. CHEN F.Y., *Kinematic synthesis of cam profiles for prescribed acceleration by a finite integration method*. Trans. ASME, J. Engng., 1973, Ind. 95B, pp. 519-524.

74.-C9. CHURCHILL F.T. and HANSEN R.S., *Theory of envelopes provide new cam-design equations*. J. Engng., 1962, 35, pp. 45-55.

75.-C10. CROSSLEY F.R.E., *How to modify positioning cams*. Machine Design, 1960, pp. 121-126.

76.-C11. CRUTCHER D.E.G., *The dynamics of valve mechanisms*. Prod. Instr. mech. Engr., 1967-68, 1, 182, Part 3L, 129.

77.-C12. CHENEY R.E., *Production of very accurate high-speed master cams*. Machinery (London), 1962, 100(2570), pp. 380-386.

78.-C13. CLAYTON J.C., *Cast Iron Camshafts in Car Production*. Design and Components in Engineering. April 1971, 16.

79.-C14. ***, *Combustion effects of asymmetric valve strategies*. In Automotive Engineering, Decembrie 1993, pp. 49-53.

80.-C15. CHOI J.K., KIM S.C., Hyundai Motor Co. Korea, *An Experimental Study on the Frictional Characteristics in the Valve Train System*. (945046), In FISITA CONGRESS, 17-21 October 1994, Beijing, pp. 374-380.

81.-C16. ***, Chrysler's *Vlo light-truck engine*. In revista Automotive Engineering, Decembrie 1993, pp. 55-57.

82.-C17. COMĂNESCU Adr., COMĂNESCU D., GEORGESCU L., *Bazele analizei şi sintezei mecanismelor cu memorie rigidă*, Edit. Politehnica Press, Bucureşti, 175 pag., 2008.

83.-D1. DRANGA M., *Contributii la analiza dinamică a mecanismelor cu unul si cu mai multe grade de mobilitate*. Teză de doctorat. I.P.B., Bucuresti, 1975.

84.-D2.   DUDITA FL., *Teoria mecanismelor*. Universitatea Brasov, 1979.

85.-D3. DEMIAN T., s.a., *Mecanisme de mecanică fină*. Editura Didactică si Pedagogică, Bucuresti, 1982.

86.-D4. DRANGA M., *Mecanisme si organe de masini, partea I. Transmisii mecanice*. I.P.B., Bucuresti, 1983.

87.-D5. DARABONT AL., s.a., *Socuri si vibratii- Aplicatii în tehnică*. Editura tehnică, Bucuresti, 1988.

88.-D6. DARABONT AL., VAITEANU D., *Combaterea poluării sonore si a vibratiilor*. Editura tehnică, Bucuresti, 1975.

89.-D7. DECIU E.,s.a., *Probleme de vibratii mecanice*. I.P.B.,Bucuresti, 1978.

90.-D8. DODESCU GH., *Metode numerice în algebră*. Editura tehnică, Bucuresti, 1979.

91.-D9. DRANGA M., *Asupra echilibrării unei structuri de robot 6R*. In al VI-lea SYROM'93, Vol. II., Bucuresti, iunie 1993.

92.-D10. DRANGA M., *Metodă de echilibrare a unui lant cinematic plan articulat*. In al IV-lea SYROM'85. Vol. III-1., Bucuresti, iulie 1985.

93.-D11. DUCA C., *Sinteza mecanismelor cu came în functie de raza de curbură a profilului*. In al IV-lea SYROM'85, Vol. III-1., Bucuresti, iulie 1985.

94.-D12. DRAGHICI I., s.a., *Suspensii si amortizoare*. E.T. , Bucuresti, 1970.

95.-D13. DUDLEY W.M., *New Methods in Valve Cam Design*. Trans. SAE, January 1948, 2, pp. 19-33.

96.-D14. DRUCE G., *Research in cam mechanisms*. I. Mech. E. Discussion on Mechanisms, 1971, 4-13.

97.-E1. ERMAN A.G., SANDOR G.N., *Kineto-elastodynamic- a review of the state of the art and rends*. Mechanism and Machine Theory nr.1., 1972.

98.-E2. EISS N.S., *Vibration of cams having tow degrees-of-fredom*. Trans. ASME, J. Engng., Ind. 86B, 1964, pp. 343-350.

99.-E3. ERISMAN R.J., *Automotive cam profile synthesis and valve gear dynamic from domensionless analysis*. Trans. SAE, 75, 1967, pp. 128-147.

100.-F1. FAWCETT G.F., FAWCETT J.N., *Comparison of polydyne and non polydyne cams*. In, Cams and cam mechanisms, Edited by J. REES JONED, MEP, London and Birmingham, Alabama, 1974.

101.-F2. FRATILA G., PETRESCU FL., s.a., *Cercetări privind transmisibilitatea vibratiilor motorului la cadrul si caroseria automobilului*. In, CONAT, Brasov, 1982.

102.-F3. FRATILA G., PETRESCU FL., s.a., *Contributii privind ameliorarea suspensiei grupului motopropulsor*. Buletinul Universitătii Brasov, 1986.

103.-F4. FENTON R.G., *Determining minimum cam size*. In Machine Design, 1966, 38(2), pp. 155-158.

104.-F5. FENTON R.G., *Cam design-determining of the minimum base radius for disc cams with reciprocating flat faced followers*. In Automobile Enginer, 3, 1967, pp. 184-187.

105.-G1. GRECU B., CANDREA A., COLTOFEANU N., *Determinarea reactiunilor dinamice în cuplele cinematice la mecanismele plane cu ajutorul modulelor de calcul*. In al VII-lea Simpozion national de roboti industriali si mecanisme spatiale. Vol. 3., Bucuresti, octombrie 1987.

106.-G2. GHITA E., *Proiectarea camelor bilaterale poliracordate*. In PRASIC'94, Brasov, decembrie 1994.

107.-G3.   GRUNWALD B., *Teoria,calculul si constructia motoarelor pentru autovehicule rutiere*. Editura didactică si pedagogică, Bucuresti, 1980.

108.-G4. GIORDANA F., s.a., *On the influence of measurement errors in the Kinematic analysis of cam*. Mechanism and Machine Theory 14 (1979), nr. 5., pp, 327-340, 1979.

109.-G5. GRADU M., *Stadiul actual al cercetărilor în domeniul mecanismelor de distributie ale motoarelor cu ardere internă*. Referat I pentru doctorat, I.P.B., Bucuresti, 1991.

110.-G6. GRUMAZESCU M., s.a., *Combaterea zgomotului si vibratiilor*. E.T., Bucuresti, 1964.

111.-G7. GAGNE A.F., *Design high speed cams*. In Machine Design, 25, 1953, pp. 121-135.

112.-G8. GRANT B., s.a., *Cam design survey*. Design Technology Transfer, ASME, 1974, pp. 177-219.

113.-G9. GRODZINSKI P., *Production of cam profiles by positive mechanisms*. Machinery (London), 1959, 88(2269), pp. 683-688.

114.-G10. GOODMAN T.P., *Linkages vs cams*. Machine Design, 1958, 30(17), pp. 102-109.

115.-G11. GRECU B., PETRESCU, F., s.a., *Mecanisme Plane – lucrări pentru laborator si proiect.* Editura BREN, Bucuresti, ISBN 978-973-648-697-5, 191 pag., 2007.

116.-H1. HANDRA-LUCA V., *Organe de masini si mecanisme*. Editura Did. si pedagogică, Bucuresti, 1975.

117.-H2. HANDRA-LUCA V.,STOICA A., *Introducere în teoria mecanismelor*. Vol. II., Editura Dacia, Cluj-Napoca, 1983.

118.-H3. HERRMANN R., DELANGE J., LOURDOUR G., *Evolution du trasee des cames*. Ingenieurs de l'automobile, nr. 11, 1969.

119.-H4. HAIN K., *Optimization of a cam mechanism to give goode transmissibility maximal output angle of swing and minimal acceleration*. Journal of Mechanisms 6 (1971), Nr. 4., pp.419-434.

120.-H5. HARRIS M.C., CREDE E.C., *Socuri si vibratii*. Vol. I-III., E.T., Bucuresti, 1968-69.

121.-H6. HEBELER C.B., *Design equation and graphs for finding the dynamic response of cycloidal-motion cam systems*. Machine Design, Feb. 1961, pp. 102-107.

122.-H7. HRONES J.A., *An analysis of Dynamic Forces in a Cam-Driver System*, Trans. ASME, 1948, 70, PP. 473-482.

123.-H8. HIRSCHHORN J., *Disc-cam curvature*. In Machine Design 31(3), 1959, pp. 125-129.

124.-H9. HALE F.W., *Cam machining without master former*. Tool Engineer, 1955, 35(6), pp. 82-87.

125.-H10. HOSAKA T., and HAMAZAKI M., *Development of the Variable Valve Timing and Lift (VTEC) Engine for the Honda NSX*, (910008), Fifth Auto-technologies Conference Proceedings, SAE,pp. 238.

126.-H11. HOORFAR, M., NAJJARAN, H., CLEGHORN, W.L, *Software demonstration of disc cam mechanisms for mechanical engineering education,* Journal: The International Journal of Mechanical Engineering Education, ISSN: 0306-4190, Volume 35 Issue 2, April 2007, pp. 166-180.

127.-I1. IACOB C., *Mecanica teoretică*. E.D.P., Bucuresti, 1971.

128.-I2. IUDIN E., s.a., *Issledovanie suma ventileatornîh ustanovok I metodov borbî s nim*. Oborongiz, Moskva, 1958.

129.-J1. JIANG QI , XU ZENG-YIN, *Compounding of mechanism and analysis and synthesis of complex mechanisms*. In al IV-lea SYROM'85, Vol. III-1., Bucuresti, iulie 1985.

130.-J2. JONES J.R., REEVE J.E., *Dynamic response of cam curves based on sinusoidal segments*. In Cams and cam mechanisms, Edited by J. REES JONES, MEP, London and Birmingham, Alabama, 1974.

131.-J3. JACOBSEN and AYRE R., *Engineering Vibration*. Mc Graw- Hill Book Co. Inc., 1958.

132.-J4. JENSEN P.W., *Cam Design and Manufacture*. Industrial Press., New York, 1965.

133.-J5. JOHNSON R.C., *A rapid method for developing cam profiles having desired acceleration characteristics*. In Machine Design 27(12), 1965, pp. 129-132.

134.-J6. JELLING W., *Precision machines assure cam accuracy*. In Iron Age, 1954, 173(15), pp. 140-142.

135.-J7. JASSEN B., *Kraftschlub bei Kurventrieben*. Ind. Anz., 1966, 88, Part. I: 1906-1907; part. II: 2193-2196.

136.-K1. KOVACS FR., PERJU D., CRUDU M., *Mecanisme. Partea I-a. Analiza mecanismelor*. I.P."Traian Vuia" din Timisoara, 1978.

137.-K2. KOVACS FR., PERJU D., *Mecanisme*. I.P. "Traian Vuia" din Timisoara, 1977.

138.-K3. KOSTER M.P., *The effects of backlash and shaft flexibility on the dynamic behaviour of a cam mechanism*. In, Cams and cam mechanisms, 1974, pp. 141-146.

139.-K4. KWAKERNAAK H., *Minimum Vibration Cam Profiles*, J. Mech. Eng. Sci., 1968, 10, pp. 219-227.

140.-K5. KLOOMOK M., s.a., *Plate cam design-evaluating dynamic loads*. Prod. Engng., 27(1), 1956, pp. 178-182.

141.-K6. KLOOMOK M., MUFFLEY R.V., *Plate cam design-pressure angle analysis*. In Product Engineering, 1955, 26(5), pp. 155-160.

142.-K7. KERLE H., *How effective is the method of finite differences as regards simple cam mechanisms*. Cams and cam mechanisms, 1974, pp. 131-135.

143.-L1. LOWN G., s.a., *Survey of Investigations in to the Dynamic Behaviour of Mechanisms Contsining Links with Distributed Mass and Elasticity*. Mech. and Mach. Th., 7, 1972.

144.-L2. LEDERER P., *Dynamische synthese der ubertragungs-funktion eines Kurvengetriebes*. In, Mech. Mach. Theory ,Vol. 28., Nr.1., pp. 23-29, Printed in Great Britain, 1993.

145.-M1. MANOLESCU N.I., KOVACS FR., ORANESCU A., *Teoria mecanismelor si a masinilor*. Editura didactică si pedagogică, Bucuresti, 1972.

146.-M2. MANOLESCU N.I., MAROS D., *Teoria mecanismelor si a masinilor*. Editura tehnică, Bucuresti, 1958.

147.-M3. MANOLESCU N.I., s.a., *Probleme de teoria mecanismelor si a masinilor*. Vol. II., E.D.P., Bucuresti, 1968.

148.-M4. MAROS D., *Mecanisme*. Vol. I., I.P. Cluj-Napoca, 1980.

149.-M5. MERTICARU V., *Mecanisme si organe de masini*. I.P.Iasi, 1979.

150.-M6. MANGERON D., IRIMICIUC N., *Mecanica rigidelor cu aplicatii în inginerie*. Vol. I,II si III. Editura tehnică, Bucuresti, 1981.

151.-M7. MARUSTER ST., *Metode numerice în rezolvarea ecuatiilor neliniare*. Ed. Tehn., Bucuresti, 1981.

152.-M8. MARINA M., *Contributii la studiul optimizării distributiei motoarelor cu ardere internă în 4 timpi*. Rezumatul tezei de doctorat, Timisoara, 1978.

153.-M9. MANEA GH., *Organe de masini*. Editura Tehnică, Bucuresti, 1970.

154.-M10. MITSI S., TSIAFIS J., *Optimal synthesis of cam mechanisms*. In SYROM'93, Vol. III., pp. 155-162., Bucuresti, iunie 1993.

155.-M11. MARINA M., *Consideration on the functional compatibility of the engine distribution mechanism springs*. SYROM'97, Vol. 3., pp. 313, Bucuresti, august 1997.

156.-M12. MERCER S., *Dynamic characteristics of cam forms calculated by the digital computer*. Trans. ASME, Nov. 1958, 80, pp. 1695-1705.

157.-M13. MARINCAS D., FRATILA G., PETRESCU FL., s.a., *Rezultatele experimentale privind îmbunătătirea izolatiei fonice a cabinei autoutilitarei TV-14*. In CONAT, Brasov, 1982.

158.-M14. MOLIAN S., *The Design of Cam Mechanisms and Linkages*. Elsevier, New York, 1968.

159.-M15. MOISE V., SIMIONESCU I., ENE M., NEACŞA M., TABĂRĂ I., *Analiza mecanismelor aplicate*, Editura Printech, ISBN 978-973-718-891-5, Bucureşti, 216 pag., 2008.

160.-N1. NEKLUTIN C.N., *Designing cams for controlled inertia and vibration*. In Machine Design, June 1952, pp. 143-153.

161.-N2. NAKANISHI F., *On cam from which induce no surging in valve springs*. Report of the Aeronautical Research Institute, 220, TOKYO Imperial University, 1941, pp. 271-280.

162.-O1. OPREAN M., *Studiul interactiunii camă-arc de supapă la motoarele, cu aprindere prin scânteie, de turatie ridicată*. Teză de doctorat, I.P.B., Bucuresti, 1984.

163.-O2. OPRISAN C., POPOVICI GH., *O analiză a variatiei unghiului de presiune la mecanismele cu camă si tachet de translatie*. In PRASIC'94, Brasov, decembrie 1994.

164.-O3. OHRNBERGER G., MANN M., AUDI A.G., Germany, *The Audi 5- Valve Cylinder Head Concept*.(945004), In XXV FISITA CONGRESS, 17-21 October 1994, Beijing, pp. 36-44.

165.-P1. PELECUDI CHR., DRANGA M., *Dinamica masinilor*. I.P.B., Bucuresti, 1980.

166.-P2. PELECUDI CHR., *Bazele analizei mecanismelor*. Editura Academiei R.S.R., Bucuresti, 1967.

167.-P3. PELECUDI CHR., *Precizia mecanismelor*. Editura Academiei R.S.R., Bucuresti, 1975.

168.-P4. PELECUDI CHR., MAROS D., MERTICARU V., PANDREA N., SIMIONESCU I., *Mecanisme*. E.D.P., Bucuresti, 1985.

169.-P5. PELECUDI CHR., s.a., *Proiectarea mecanismelor*. I.P.B., Bucuresti, 1981.

170.-P6. PELECUDI CHR., s.a., *Probleme de mecanisme*. Editura didactică si pedagogică, Bucuresti, 1982.

171.-P7. PELECUDI CHR., s.a., *Algoritmi si programe pentru analiza mecanismelor*. Editura tehnică, Bucuresti, 1982.

172.-P8. PELECUDI CHR., SIMIONESCU I., ENE M., CANDREA A., STOENESCU M., MOISE V., *Mecanisme cu cuple superioare: came si roti*. I.P.B., Bucuresti, 1982.

173.-P9. POPESCU I., *Proiectarea mecanismelor plane*. Editura Scrisul Românesc din Craiova, 1977.

174.-P10. PANDREA N., MUNTEANU M., *Curs de vibratii*. Vol. I. si II., I.P.B., Bucuresti, 1979.

175.-P11. PELECUDI CHR., SAVA I., *Studiul experimental al dinamicii mecanismelor cu came*. In revista Studii si cercetări de mecanică aplicată, nr. 3., Bucuresti, 1970.

176.-P12. PELECUDI CHR., SAVA I., MATHEESCU A., *Optimizarea legilor de functionare ale mecanismelor de distributie*. In revista Studii si cercetări de mecanică aplicată, nr. 3., Bucuresti, 1968.

177.-P13. PFISTER F., FAYET M., *Linearization of dynamic models*. In al VI-lea SYROM'93, Vol. II., Bucuresti, iunie 1993.

178.-P14 PELECUDI CHR., BOGDAN R., *Sinteza mecanismelor cu came la prescrierea valorilor arcelor de curbă*. In revista Studii si cercetări de mecanică aplicată, nr. 6., Bucuresti, 1962.

179.-P15. PELECUDI CHR., MATHEESCU A., *Analiza armonică a legilor de miscare la mecanismele cu camă*. In revista Studii si cercetări de mecanică aplicată, nr. 1., Bucuresti, 1969.

180.-P16. PELECUDI CHR., SAVA I., *Asupra analizei si sintezei mecanismelor cu came*. In revista Constructia de masini, nr. 8-9., Bucuresti, 1967.

181.-P17. PANDREA N., HARA V., POPA D., *Sinteza dimensională a mecanismelor de distributie cu admisie adaptivă pentru optimizarea legii de deplasare a supapei de admisie*. In PRASIC'94, Brasov, dec. 1994.

182.-P18. POPOVICI GH., *Sinteza profilului camei cu tachet de translatie*. In PRASIC'94, Brasov, decembrie 1994.

183.-P19. POPOVICI GH., LEOHCHI D., CIAUSU V., *Sinteza profilului camei cu tachet oscilant*. In PRASIC'94, Brasov, dec. 1994.

184.-P20. PELECUDI CHR., SAVA I., *Optimizări în sinteza numerică a miscării mecanismelor cu came*. In revista Studii si cercetări de mecanică aplicată, nr. 5., Bucuresti, 1971.

185.-P21. PETRESCU F., PETRESCU R., *Contributii la optimizarea legilor polinomiale de miscare a tachetului de la mecanismul de distributie al motoarelor cu ardere internă*. In E.S.F.A.'95, Vol. 1.,pp. 249-256., Bucuresti, mai 1995.

186.-P22. PETRESCU F., PETRESCU R., *Contributii la sinteza mecanismelor de distributie ale motoarelor cu ardere internă*. In E.S.F.A.'95, Vol. 1., pp. 257-264., Bucuresti, mai 1995.

187.-P23. PETRESCU F., PETRESCU V., *Dinamica mecanismelor cu came (exemplificată pe mecanismul clasic de distributie)*. SYROM'97, Vol. 3., pp. 353-358., Bucuresti, august 1997.

188.-P24. PETRESCU F., PETRESCU V., *Contributii la sinteza mecanismelor de distributie ale motoarelor cu ardere internă cu metoda coordonatelor carteziene*. SYROM'97, Vol. 3., pp. 359-364., Bucuresti, august 1997.

189.-P25. PETRESCU F., PETRESCU V., *Contributii la maximizarea legilor polinomiale pentru cursa activă a mecanismului de distributie de la motoarele cu ardere internă*. SYROM'97, Vol. 3., pp. 365-370., Bucuresti, august 1997.

190.-P26. PETRESCU F.,PETRESCU V., *Sinteza mecanismelor de distributie prin metoda coordonatelor rectangulare (carteziene)*. In Conferinta "Grafica-2000", Universitatea din Craiova, Craiova, 2000.

191.-P27. PETRESCU F., PETRESCU V., *Designul (sinteza) mecanismelor cu came prin metoda coordonatelor polare (metoda triunghiurilor)*. In Conferinta "Grafica-2000", Universitatea din Craiova, Craiova, 2000.

192.-P28. PETRESCU F., PETRESCU V., *Legi de mişcare pentru mecanismele cu came*. In al VII-lea Simpozion Naţional cu Participare Internaţională Proiectarea Asistată de Calculator, PRASIC'02, Braşov, 2002, Vol. I, p. 321-326.

193.-P29. PETRESCU, F., PETRESCU, R. *Elemente de dinamica mecanismelor cu came*. In al VII-lea Simpozion Naţional cu Participare Internaţională Proiectarea Asistată de Calculator, PRASIC'02, Braşov, 2002, Vol. I, p. 327-332.

194.-P30. PETRESCU, V., PETRESCU, I., ANTONESCU, O. *Randamentul cuplei superioare de la angrenajele cu roţi dinţate cu axe fixe*. In al VII-lea Simpozion Naţional cu Participare Internaţională Proiectarea Asistată de Calculator, PRASIC'02, Braşov, 2002, Vol. I, p. 333-338.

195.-P31. PETRESCU, I., PETRESCU, V., OCNĂRESCU, C. *The Cam Synthesis With Maximal Efficiency*. In al VII-lea Simpozion Naţional cu Participare Internaţională Proiectarea Asistată de Calculator, PRASIC'02, Braşov, 2002, Vol. I, p. 339-344.

196.-P32. PETRESCU, F., PETRESCU, R. *Câteva elemente privind îmbunătăţirea designului mecanismului motor*. În al VIII-lea Simpozion Naţional, de Geometrie Descriptivă, Grafică Tehnică şi Design, GTD 2003, Braşov, iunie 2003, Vol. I, p. 353-358.

197.-P33. PETRESCU, F., PETRESCU, R. *The cam design for a better efficiency*. In the International Conference on Engineering Graphics and Design, ICEGD 2005, Bucharest, 2005, Vol. I, p. 245-248.

198.-P34. PETRESCU, F.I., PETRESCU, R.V. *Contributions at the dynamics of cams*. In the Ninth IFToMM International Symposium on Theory of Machines and Mechanisms, SYROM 2005, Bucharest, Romania, 2005, Vol. I, p. 123-128.

199.-P35. PETRESCU, F.I., PETRESCU, R.V. *Determining the dynamic efficiency of cams*. In the Ninth IFToMM International Symposium on Theory of Machines and Mechanisms, SYROM 2005, Bucharest, Romania, 2005, Vol. I, p. 129-134.

200.-P36. PETRESCU, F.I., PETRESCU, R.V. *An original internal combustion engine*. In the Ninth IFToMM International Symposium on Theory of Machines and Mechanisms, SYROM 2005, Bucharest, Romania, 2005, Vol. I, p. 135-140.

201.-P37. PETRESCU, R.V., PETRESCU, F.I. *Determining the mechanical efficiency of Otto engine's mechanism*. In the Ninth IFToMM International Symposium on Theory of Machines and Mechanisms, SYROM 2005, Bucharest, Romania, 2005, Vol. I, p. 141-146.

202.-P38. PETRESCU, F.I., PETRESCU, R.V., POPESCU N., *The efficiency of cams*. In the Second International Conference "Mechanics and Machine Elements", Technical University of Sofia, November 4-6, 2005, Sofia, Bulgaria, Vol. II, p. 237-243.

203.-R1. RADOI M., DECIU E., *Mecanica*. E.D.P., Bucuresti, 1973.

204.-R2. RADOI M., DECIU E., *Mecanica*. E.D.P., Bucuresti, 1977.

205.-R3. RAO A., *Optimum Elastodynamic Synthesis of a Cam-Follower Train Using Stochastic-Geometric Programming*. Mech. and Mach. Theory, Vol. 15., 1980.

206.-R4. RAICU A., *Consideratii privind nedeterminarea din ecuatia de miscare a masinii*. In PRASIC, Brasov, decembrie 1994.

207.-R5. REES JONES J., *Analog simulation of SCCA cam motion*. In Mech. Eng. Deptl. Report, 1974, Liverpool Polytechnic.

208.-R6. ROSKILLY M., s.a., *Valve gear design analysis*. In XXII FISITA CONGRESS (865027), PP. 1.193-1.200.

209.-R7. ***, Revue Technique, aprilie 1991, pp. 22.

210.-S1. SILAS GH., *Mecanică-vibratii mecanice*, E.D.P., Bucuresti, 1968.

211.-S2. SILAS GH., s.a., *Culegere de probleme de vibratii mecanice*. Editura tehnică, Bucuresti, 1967.

212.-S3. SARSTEN A.,VALLEND H., *Computer aided design of valve cams*. Internal Combustion Engines conference, Bucharest, Paper II-19, 1967.

213.-S4. SAVA I., *Stadiul actual în dinamica mecanismelor cu came*. I-II., Rev. S.C.M.A., Nr. 5., 1969.

214.-S5. SAVA I., *Contributii la dinamica si sinteza optimală a mecanismelor cu came*. Teză de doctorat, I.P.B., 1970.

215.-S6. SAVA I., *Cu privire la functionarea in regim dinamic a supapei mecanismului distributiei motoarelor cu ardere interna*. In revista C.M. Nr.12.,Bucuresti, 1971.

216.-S7. SAVIUC S., *Optimizarea duratei de deschidere simultană a supapelor la motoarele cu aprindere prin scânteie*. Teză de doctorat, I.P.B., 1979.

217.-S8. SIRETEANU T., GRUNDISCH O., PARAIAN S., *Vibratiile aleatoare ale automobilelor*. E.T., Bucuresti, 1981.

218.-S9. STOICESCU A., *Dinamica autovehiculelor*. Vol. I-II., I.P.B., Bucuresti, 1980-82.

219.-S10. STOICESCU A., *Dinamica autovehiculelor pe roti*. E.D.P., Bucuresti, 1981.

220.-S11. SONO H., UMIYAMA H., Honda RDCo., Ltd. Japan, *A study of Combustion Stability of Non-Throttling S.I. Engine with Early Intake Valve Closing Mechanism*. (945009), In XXV FISITA CONGRES, October 1994, Beijing, pp. 78-87.

221.-T1. TEMPEA I., POPA GH., *Mecanisme plane articulate*. I.P.B., Bucuresti, 1978.

222.-T2. TEMPEA I., MARTINEAC A., *Organe de masini, teoria mecanismelor si prelucrării prin aschiere. Partea I , mecanisme*, I.P.B., Bucuresti, 1983.

223.-T3. TEMPEA I., BALESCU C., ADIR G., *Mecanism de presare destinat mecanizării operatiei de formare în rame (părtile I si II)*. In al VII-lea Simpozion national de roboti industriali si mecanisme spatiale. Vol. 3., Bucuresti, 1987.

224.-T4. TEMPEA I., GRADU M., *Sinteza camei de translatie cu tachet cu rolă, cu ajutorul functiilor spline*. In lucrările simpozionului de R.I., Timisoara, 1992.

225.-T5. TUTUNARU D., *Mecanisme plane rectiliniare si inversoare*. Editura tehnică, Bucuresti, 1969.

226.-T6. TORAZZA G., *A variable lift and event control device piston engine valve operation*. In FISITA XIV Congres, Paper II / 10, London, 1972.

227.-T7. TESAR D., MATTHEW G.K., *The design of modelled cam sistems*. In Cams and cam mechanisms, 1974.

228.-T8. TERME D., *Besondere Merkmalebeider Nutzung des Pressungwinkels fur kurvengetriebeanalyse und-Synthese*. In SYROM'85, Vol. III-2, pp. 489-504, Bucuresti, iulie 1985.

229.-T9. TEMPEA I., DUGĂEȘESCU I., NEACȘA M., *Mecanisme. Noțiuni teoretice și teme de proiect rezolvate*, Ed. Printech, ISBN (10) 973-718-560-9, 2006.

230.-T10. D. Taraza, N.A. Henein, W. Bryzik, "The Frequency Analysis of the Crankshaft's Speed Variation: A reliable Tool for Diesel Engine Diagnosis," *ASME Journal for Gas Turbines and Power* 123(2), 428-432, 2001

231.-T11. D. Taraza, "Accuracy Limits of IMEP Determination from Crankshaft Speed Measurements," *SAE Transactions, Journal of Engines* 111, 689-697, 2002.

232.-T12. D. Taraza, "Statistical Correlation Between the Crankshaft's Speed Variation and Engine Performance, Part I: Theoretical Model," *ASME Journal of Engineering for Gas Turbines and Power* 125(3), 791-796, 2003.

233.-T13. D. Taraza, "Statistical Correlation Between the Crankshaft's Speed Variation and Engine Performance, Part II: Detection of Deficient Cylinders and MIP Calculation," *ASME journal of Engineering for Gas Turbines and Power* 125(3), 797-803, 2003.

234.-U1. ULF A., WILLIAM S., *A Simple Procedure for Modifying High-Speed Cam Profiles for Vibration Reduction,* Journal of Mechanical Design - November 2004 - Volume 126, Issue 6, pp. 1105-1108.

235.-V1. VOINEA R., VOICULESCU D., CEAUSU V., *Mecanica*. E.D.P., Bucuresti, 1975.

236.-V2. VOINEA R., ATANASIU M., *Metode analitice noi în teoria mecanismelor*. Editura tehnică, Bucuresti, 1964.

237.-V3. Van de Straete, H.J., De Schutter, J., *Hybrid cam mechanisms*, Mechatronics, IEEE/ASME Transactions on Volume 1, Issue 4, Dec. 1996 Page(s):284 - 289

238.-W1. WIEDERRICH J.L., ROTH B., *Design of low vibration cam profiles*. In Cams and cam mechanisms, Edited by J. REES JONES, MEP, London and Birmingham, Alabama, 1974.

239.-W2. WIEDERRICH J.L., ROTH B., *Dynamic Synthesis of Cams Using Finite Trigonometric Series*, Trans. ASME, 1974.

240.-Y1. YOUNG V.C., *Considerations în valve gear design*. Trans. SAE, 1, 1947, pp. 359-365.

241.-Z1. ZHANG J.L., LI Z., *Research on the dynamics of a RSCR spatial mechanisms considering bearing clearances*. In al VI-lea SYROM, Vol. II, Bucuresti, iunie 1993.

Annex

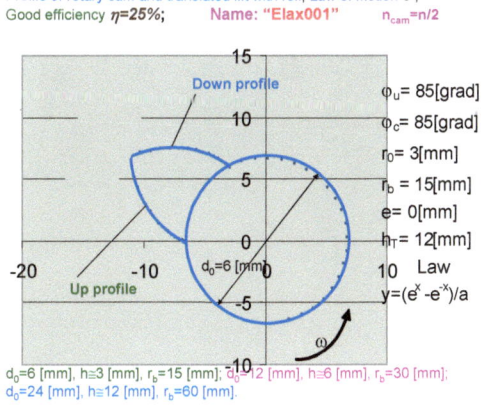

Profile of rotary cam and translated lift with roll; Law of motion e^x;
Good efficiency $\eta=25\%$;   Name: "Elax001" $n_{cam}=n/2$

φ_u= 85[grad]
φ_c= 85[grad]
r_0= 3[mm]
r_b = 15[mm]
e= 0[mm]
h_T= 12[mm]
Law
$y=(e^x-e^{-x})/a$

Down profile
Up profile
d_0=6 [mm]

d_0=6 [mm], h≅3 [mm], r_b=15 [mm]; d_0=12 [mm], h≅6 [mm], r_b=30 [mm]; d_0=24 [mm], h≅12 [mm], r_b=60 [mm].

Dynamic analysis to the rotary cam and translated follower with roll;
Law of motion e^x;   **Name: "Elax001"**
Proposed crankshaft rotation speed n=5500 [r/m]; n_{cam}=n/2

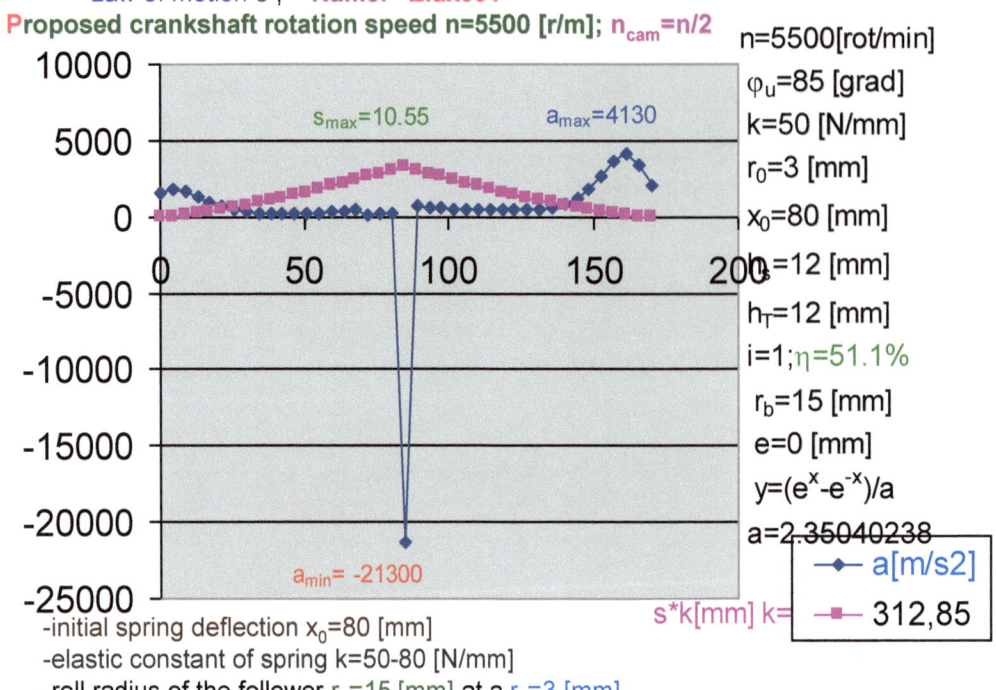

n=5500[rot/min]
φ_u=85 [grad]
k=50 [N/mm]
r_0=3 [mm]
x_0=80 [mm]
h_s=12 [mm]
h_T=12 [mm]
i=1; η=51.1%
r_b=15 [mm]
e=0 [mm]
$y=(e^x-e^{-x})/a$
a=2.35040238

- s_{max}=10.55
- a_{max}=4130
- a_{min}= -21300

-initial spring deflection x_0=80 [mm]
-elastic constant of spring k=50-80 [N/mm]
-roll radius of the follower r_b=15 [mm] at a r_0=3 [mm].

a[m/s2]
s*k[mm] k=   312,85

Profile of rotary cam and translated lift with roll; Law of motion e^x;
Best efficiency η=51%;   **Name: "2Elax001"**   **![n_{cam}=n/4]!**

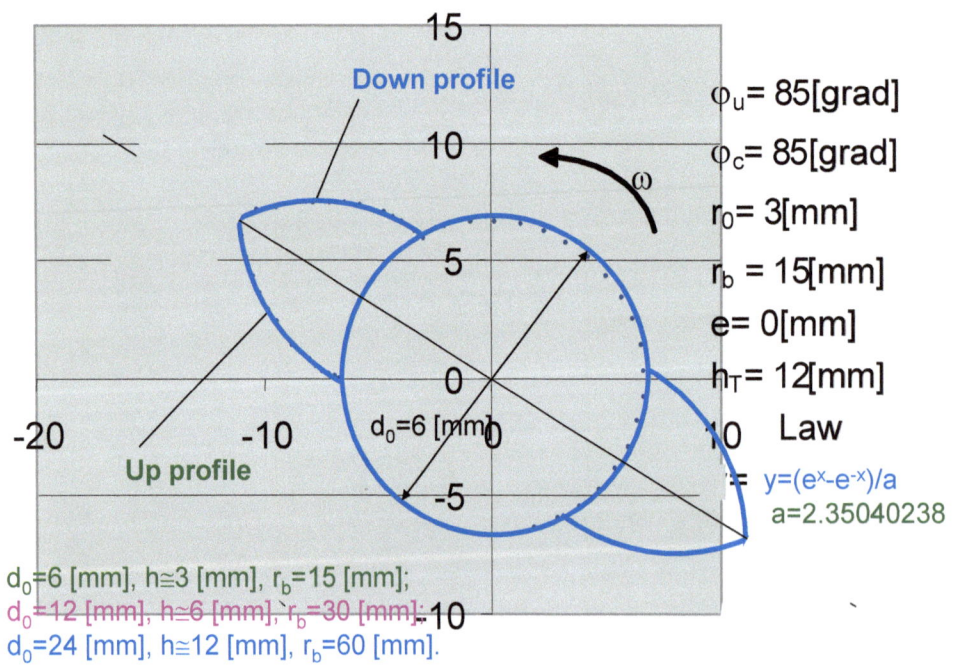

Down profile
Up profile

φ_u= 85[grad]
φ_c= 85[grad]
r_0= 3[mm]
r_b = 15[mm]
e= 0[mm]
h_T= 12[mm]
Law
$y=(e^x-e^{-x})/a$
a=2.35040238

d_0=6 [mm]
ω

d_0=6 [mm], h≅3 [mm], r_b=15 [mm];
d_0=12 [mm], h≅6 [mm], r_b=30 [mm];
d_0=24 [mm], h≅12 [mm], r_b=60 [mm].

Profile of rotary cam and translated lift with roll; Law of motion Log Nat;
Good efficiency η=25-33%;   Name: "2LogNat001"   ![n_cam=n/4]!

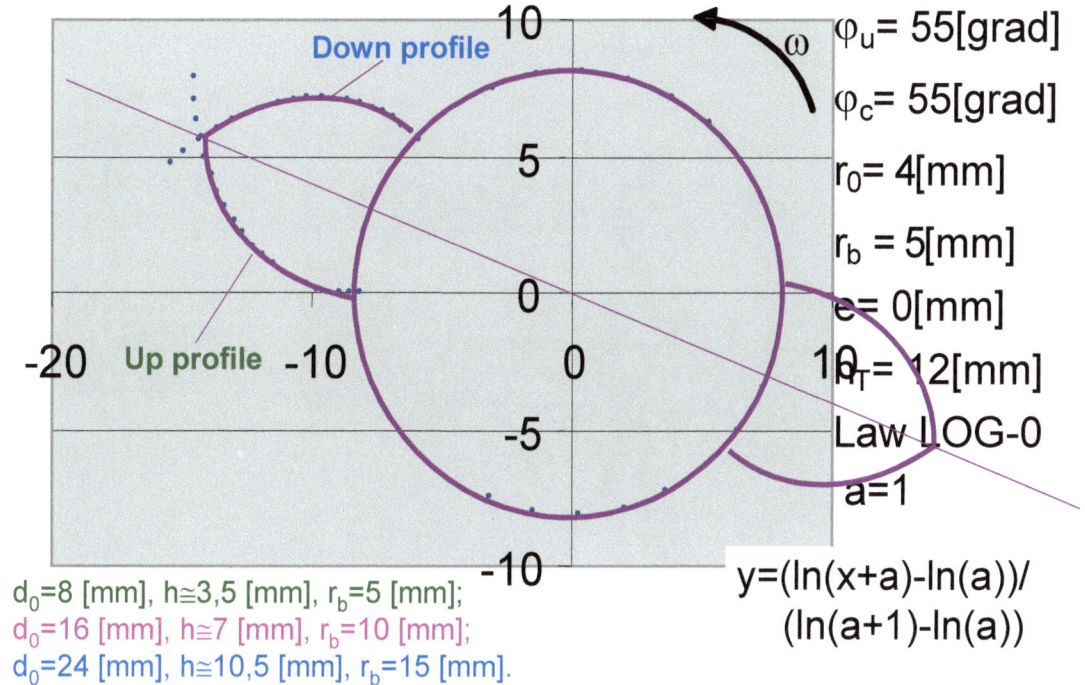

φ_u= 55[grad]

φ_c= 55[grad]

r_0= 4[mm]

r_b = 5[mm]

e= 0[mm]

h_T= 12[mm]

Law LOG-0

a=1

$y=(\ln(x+a)-\ln(a))/(\ln(a+1)-\ln(a))$

d_0=8 [mm], h≅3,5 [mm], r_b=5 [mm];
d_0=16 [mm], h≅7 [mm], r_b=10 [mm];
d_0=24 [mm], h≅10,5 [mm], r_b=15 [mm].

Dynamic analysis to the rotary cam and translated lift with roll; Law of motion Log Nat;
Good efficiency η=25-33%; Name: "2LogNat001" , ![n_cam=n/4]!

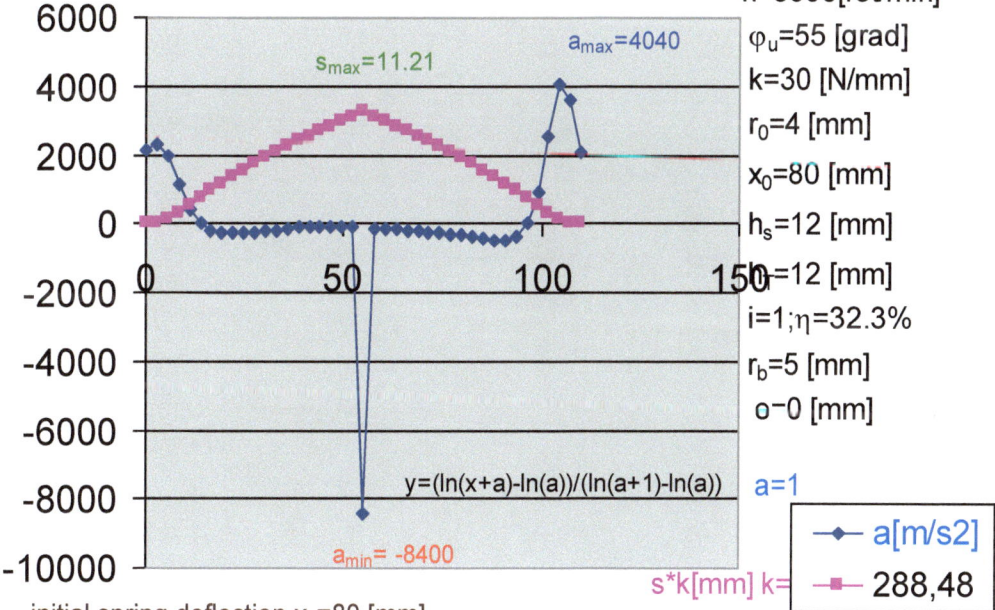

n=5500[rot/min]
φ_u=55 [grad]
k=30 [N/mm]
r_0=4 [mm]
x_0=80 [mm]
h_s=12 [mm]
h=12 [mm]
i=1;η=32.3%
r_b=5 [mm]
o−0 [mm]

a=1

a_max=4040
s_max=11.21
a_min= -8400
y=(ln(x+a)-ln(a))/(ln(a+1)-ln(a))

s*k[mm] k=

	a[m/s2]
	288,48

-initial spring deflection x_0=80 [mm]
-elastic constant of spring k=30-60 [N/mm]
-roll radius of the follower r_b=5 [mm] at a r_0=4 [mm].

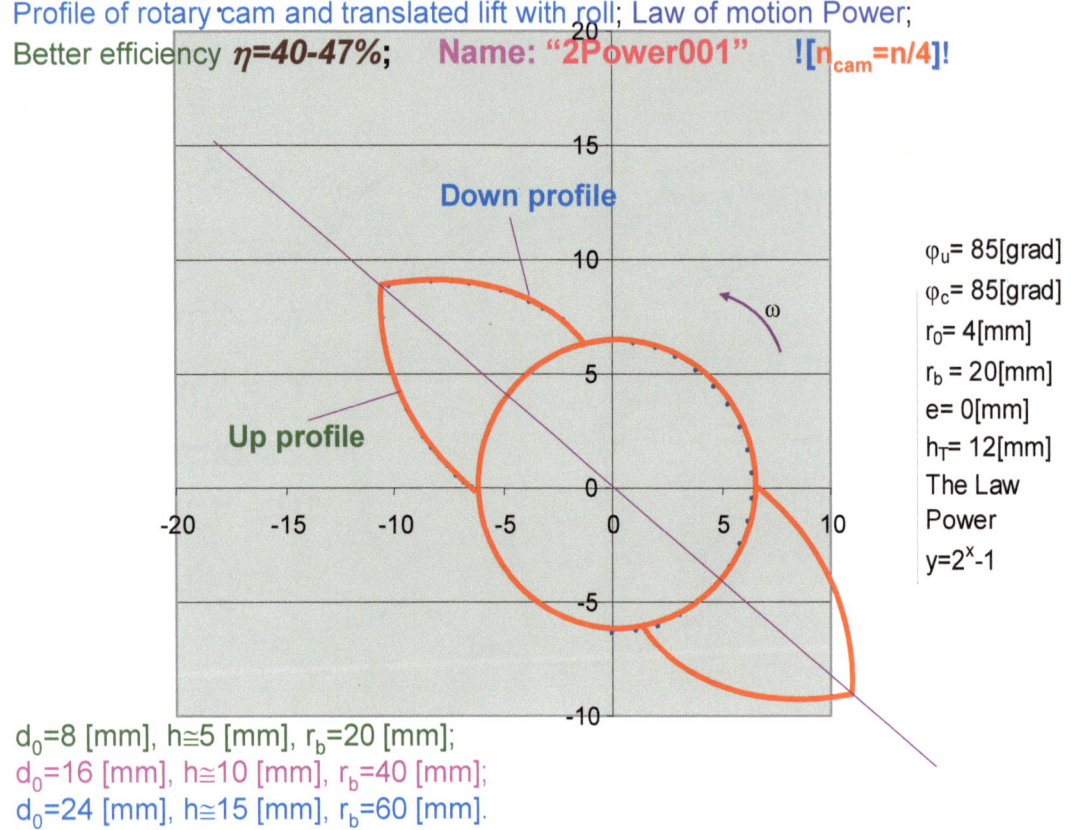

Profile of rotary cam and translated lift with roll; Law of motion Power;
Better efficiency η=40-47%;   Name: "2Power001"   ![n_{cam}=n/4]!

Down profile

Up profile

ω

φ_u= 85[grad]
φ_c= 85[grad]
r_0= 4[mm]
r_b = 20[mm]
e= 0[mm]
h_T= 12[mm]
The Law
Power
$y=2^x-1$

d_0=8 [mm], h$\cong$5 [mm], r_b=20 [mm];
d_0=16 [mm], h$\cong$10 [mm], r_b=40 [mm];
d_0=24 [mm], h$\cong$15 [mm], r_b=60 [mm].

Dynamic analysis to the rotary cam and translated lift with roll; Law of motion Power;
Better efficiency η=40-47%; Name: "2Power001" ![n_{cam}=n/4]!

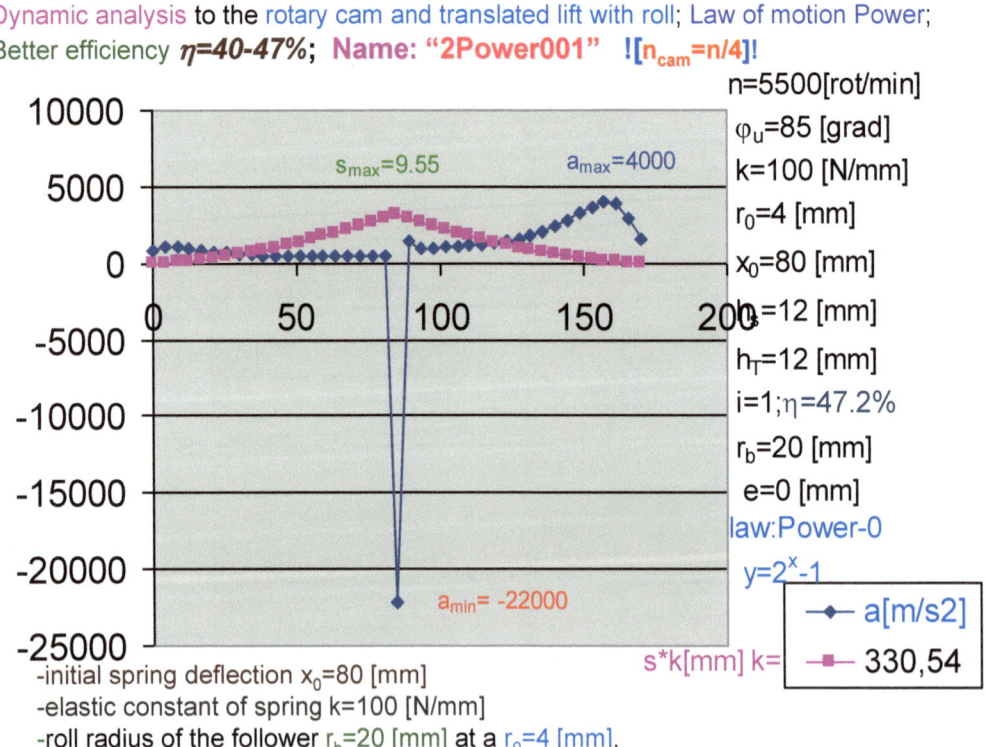

s_{max}=9.55 a_{max}=4000

a_{min}= -22000

n=5500[rot/min]
φ_u=85 [grad]
k=100 [N/mm]
r_0=4 [mm]
x_0=80 [mm]
h=12 [mm]
h_T=12 [mm]
i=1;η=47.2%
r_b=20 [mm]
e=0 [mm]
law:Power-0
$y=2^x-1$

s*k[mm] k=

a[m/s2]	
330,54	

-initial spring deflection x_0=80 [mm]
-elastic constant of spring k=100 [N/mm]
-roll radius of the follower r_b=20 [mm] at a r_0=4 [mm].

Profile of rotary cam and translated lift with roll; Law of motion Atan;
Good efficiency η=30-40%;   Name: "2Atan001"   ![n_{cam}=n/4]!

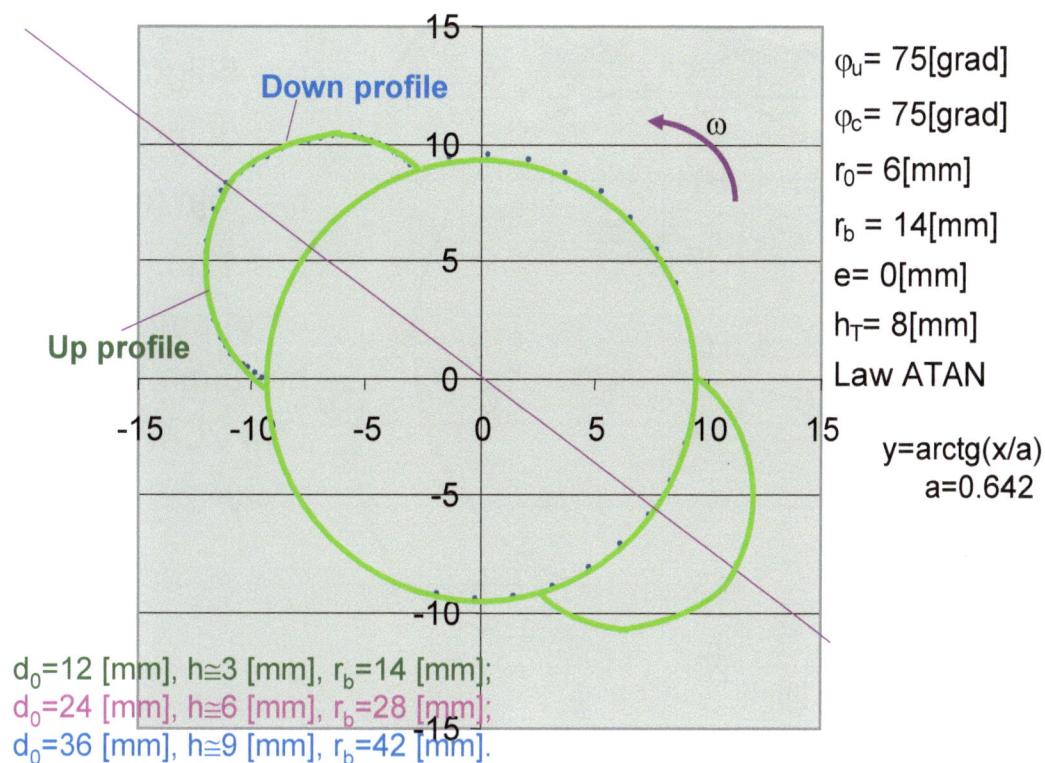

φ_u= 75[grad]
φ_c= 75[grad]
r_0= 6[mm]
r_b = 14[mm]
e= 0[mm]
h_T= 8[mm]
Law ATAN

y=arctg(x/a)
a=0.642

d_0=12 [mm], h≅3 [mm], r_b=14 [mm];
d_0=24 [mm], h≅6 [mm], r_b=28 [mm];
d_0=36 [mm], h≅9 [mm], r_b=42 [mm].

Dynamic analysis to the rotary cam and translated lift with roll; Law of motion Atan;
Good efficiency η=30-40%; Name: "2Atan001" ![n_{cam}=n/4]!

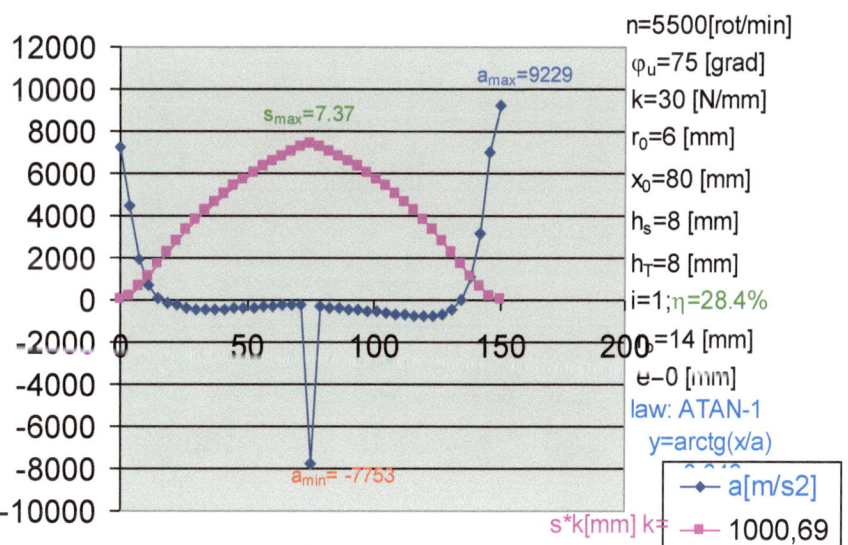

n=5500[rot/min]
φ_u=75 [grad]
k=30 [N/mm]
r_0=6 [mm]
x_0=80 [mm]
h_s=8 [mm]
h_T=8 [mm]
i=1; η=28.4%
r_b=14 [mm]
e=0 [mm]
law: ATAN-1
y=arctg(x/a)

-initial spring deflection x_0=80 [mm]
-elastic constant of spring k=30 [N/mm]
-roll radius of the follower r_b=14 [mm] at a r_0=6 [mm].

81

Profile of rotary cam and translated lift with roll; Law of motion C4P1-001;
Good efficiency η=20-35%;   Name: "2C4P1-001"   ![n_{cam}=n/4]!

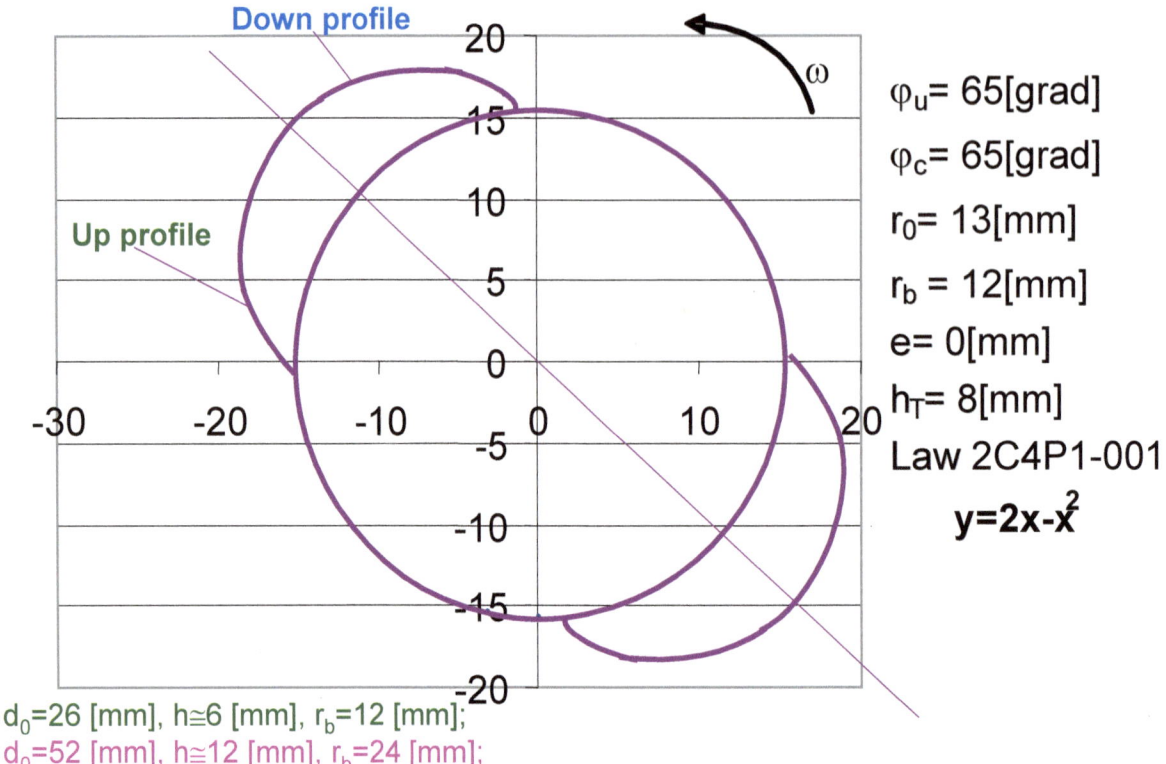

Down profile

Up profile

φ_u= 65[grad]

φ_c= 65[grad]

r_0= 13[mm]

r_b = 12[mm]

e= 0[mm]

h_T= 8[mm]

Law 2C4P1-001

$y=2x-x^2$

d_0=26 [mm], h$\cong$6 [mm], r_b=12 [mm];
d_0=52 [mm], h$\cong$12 [mm], r_b=24 [mm];

Dynamic analysis to the rotary cam and translated lift with roll; Law of motion C4P1-001;
Good efficiency η=20-35%; Name: "2C4P1-001" ![n_{cam}=n/4]!

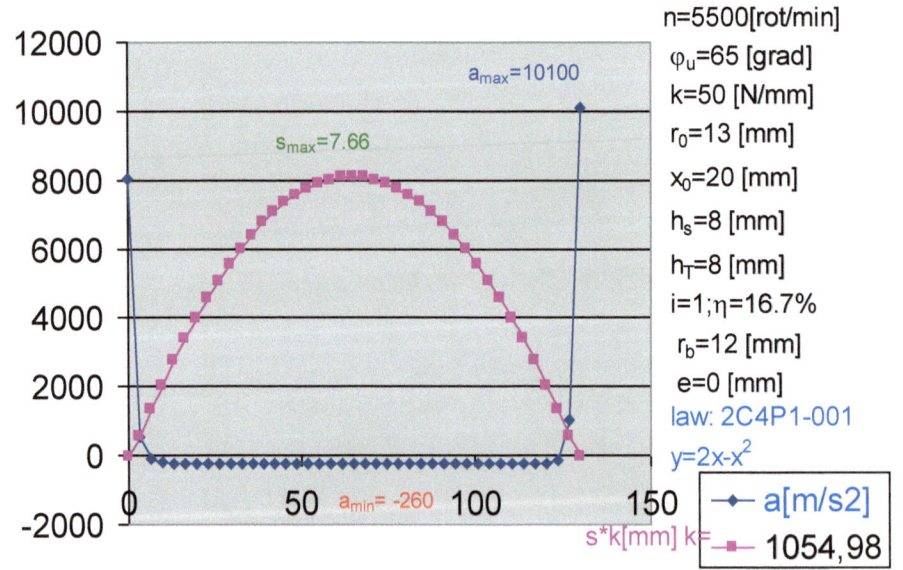

n=5500[rot/min]

φ_u=65 [grad]

k=50 [N/mm]

r_0=13 [mm]

x_0=20 [mm]

h_s=8 [mm]

h_T=8 [mm]

i=1;η=16.7%

r_b=12 [mm]

e=0 [mm]

law: 2C4P1-001

$y=2x-x^2$

a_{max}=10100

s_{max}=7.66

a_{min}= -260

s*k[mm] k=

a[m/s2]

1054,98

-initial spring deflection x_0=20 [mm]
-elastic constant of spring k=50 [N/mm]
-roll radius of the follower r_b=12 [mm] at a r_0=13 [mm].

Profile of rotary cam and translated lift with roll; Law of motion 4C4P1-001;
Best efficiency η=40-70%; Name: "4C4P1-001" ![n$_{cam}$=n/8]!

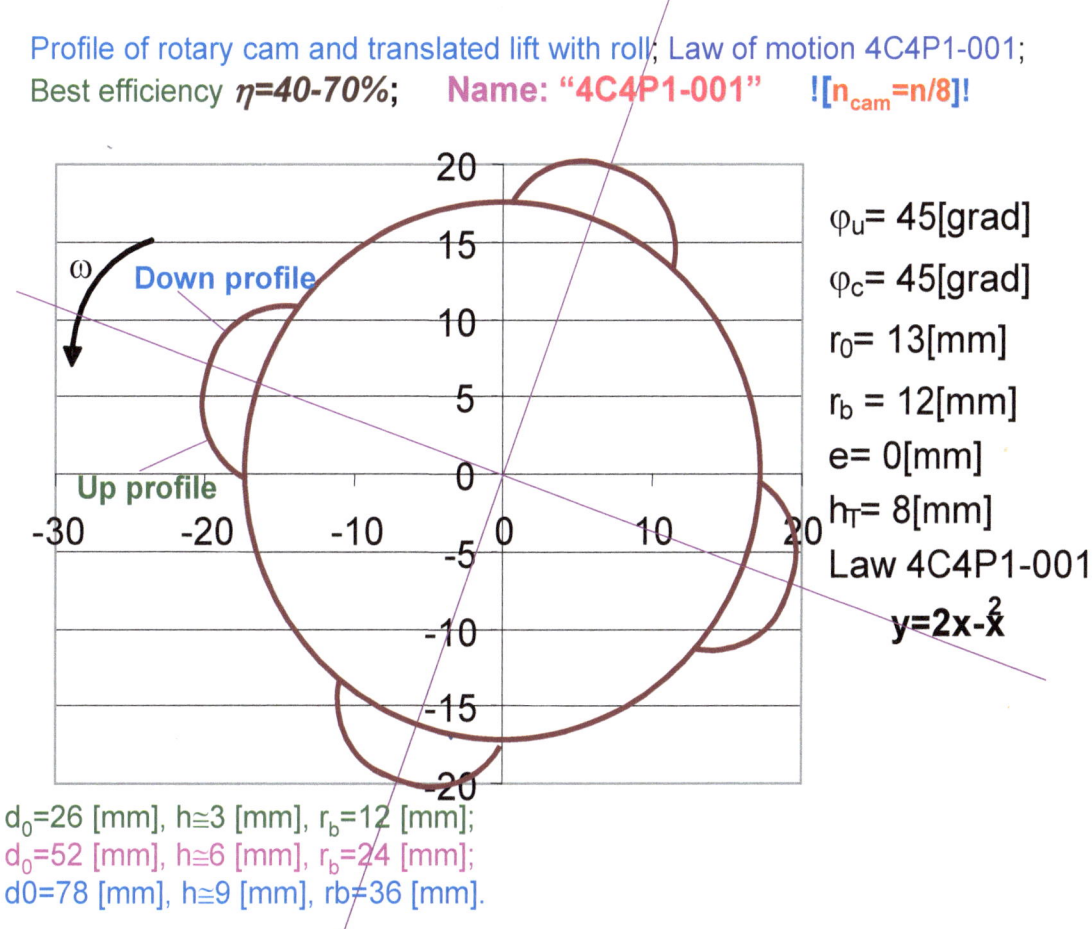

φ_u= 45[grad]

φ_c= 45[grad]

r_0= 13[mm]

r_b = 12[mm]

e= 0[mm]

h_T= 8[mm]

Law 4C4P1-001

y=2x-x^2

d_0=26 [mm], h≅3 [mm], r_b=12 [mm];
d_0=52 [mm], h≅6 [mm], r_b=24 [mm];
d0=78 [mm], h≅9 [mm], rb=36 [mm].

Dynamic analysis to the rotary cam and translated lift with roll; Law of motion C4P1-001;
Best efficiency η=40-70%; Name: "4C4P1-001" ![n$_{cam}$=n/8]!

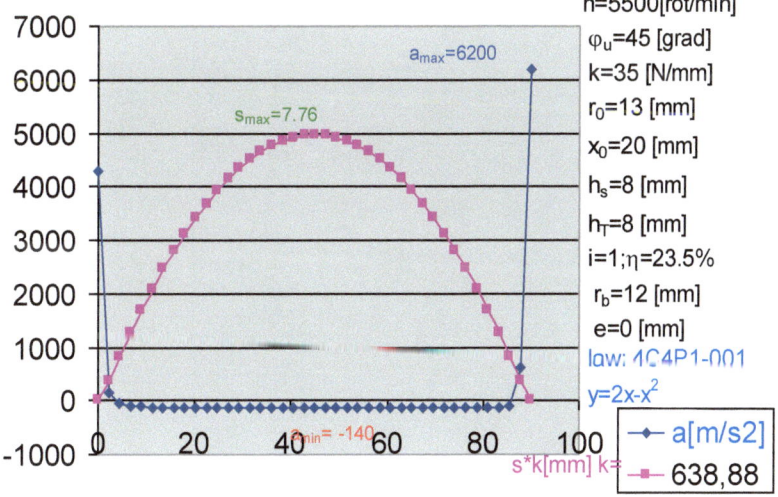

n=5500[rot/min]
φ_u=45 [grad]
k=35 [N/mm]
r_0=13 [mm]
x_0=20 [mm]
h_s=8 [mm]
h_T=8 [mm]
i=1;η=23.5%
r_b=12 [mm]
e=0 [mm]
law: 4C4P1-001
y=2x-x^2

-initial spring deflection x_0=20 [mm]
-elastic constant of spring k=35 [N/mm]
-roll radius of the follower r_b=12 [mm] at a r_0=13 [mm].

Profile of rotary cam and translated lift with roll; Law of motion SIN-001; Good efficiency η=15-30%;   Name: "2SIN-001"   ![n_{cam}=n/4]!

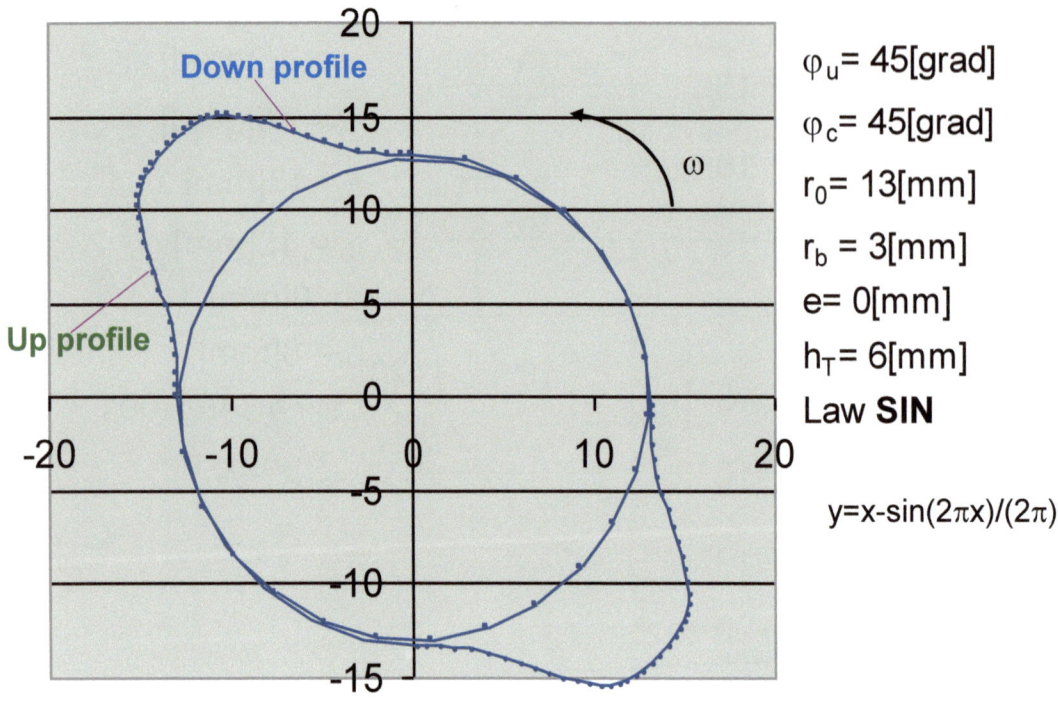

φ_u= 45[grad]

φ_c= 45[grad]

r_0= 13[mm]

r_b = 3[mm]

e= 0[mm]

h_T= 6[mm]

Law **SIN**

$y=x-\sin(2\pi x)/(2\pi)$

d_0=26 [mm], h$\cong$6 [mm], r_b=3 [mm];
d_0=52 [mm], h$\cong$12 [mm], r_b=6 [mm].

Dynamic analysis to the rotary cam and translated lift with roll; Law of motion SIN-001; Good efficiency η=15-30%; Name: "2SIN-001" ![n_{cam}=n/4]!

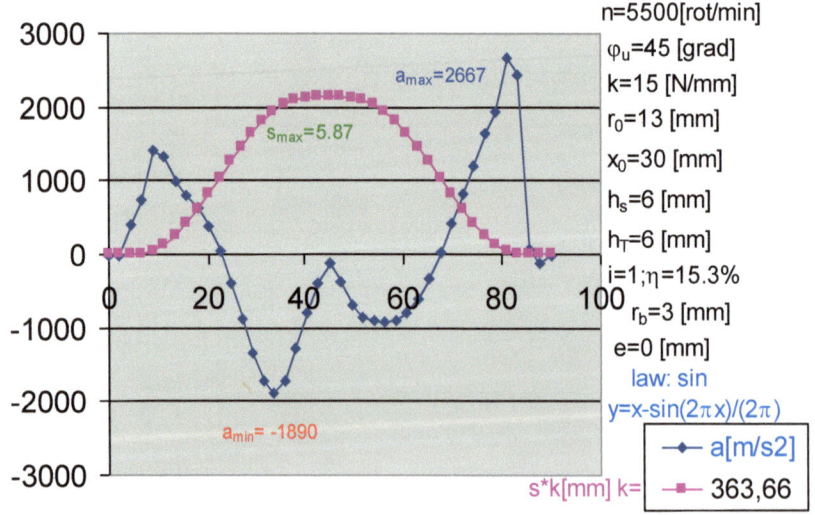

n=5500[rot/min]

φ_u=45 [grad]

k=15 [N/mm]

r_0=13 [mm]

x_0=30 [mm]

h_s=6 [mm]

h_T=6 [mm]

i=1; η=15.3%

r_b=3 [mm]

e=0 [mm]

law: sin

$y=x-\sin(2\pi x)/(2\pi)$

a[m/s2]

s*k[mm] k=   363,66

-initial spring deflection x_0=30 [mm]
-elastic constant of spring k=15 [N/mm]
-roll radius of the follower r_b=3 [mm] at a r_0=13 [mm].

Profile of rotary cam and translated lift with roll; Law of motion SIN-001;
Better efficiency η=15-60%;   Name: "4SIN-001"   ![n_{cam}=n/8]!

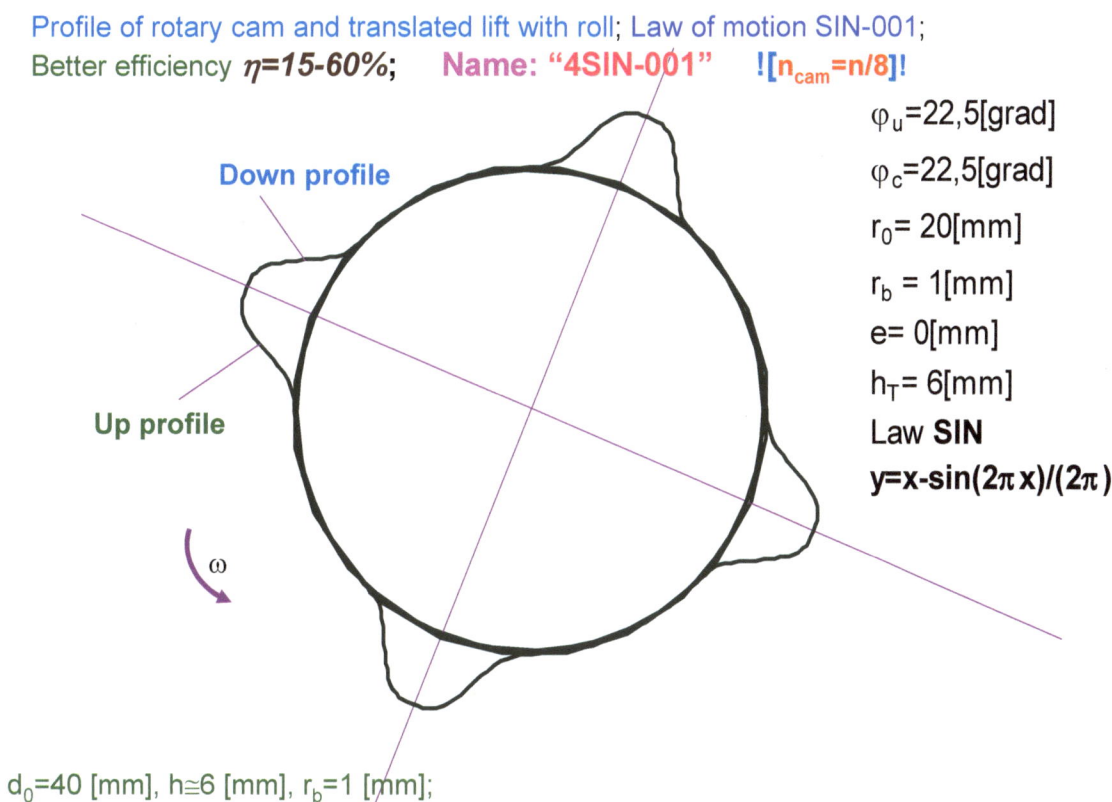

φ_u=22,5[grad]

φ_c=22,5[grad]

r_0= 20[mm]

r_b = 1[mm]

e= 0[mm]

h_T= 6[mm]

Law **SIN**

y=x-sin(2πx)/(2π)

Down profile

Up profile

ω

d_0=40 [mm], h$\cong$6 [mm], r_b=1 [mm];
d_0=60 [mm], h$\cong$9 [mm], r_b=1,5 [mm].

Dynamic analysis to the rotary cam and translated lift with roll; Law of motion SIN-001;
Better efficiency η=15-60%; Name: "4SIN-001"   ![n_{cam}=n/8]!

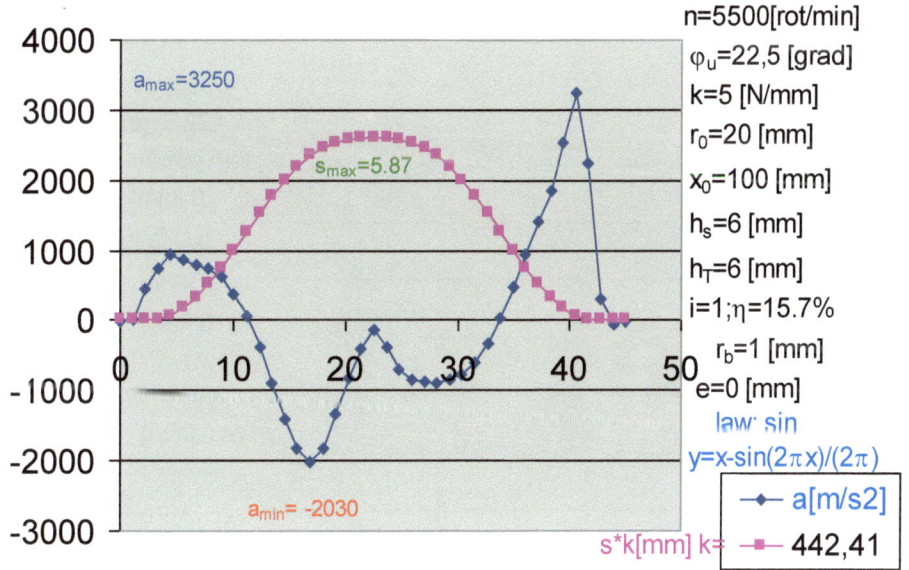

n=5500[rot/min]

φ_u=22,5 [grad]

k=5 [N/mm]

r_0=20 [mm]

x_0=100 [mm]

h_s=6 [mm]

h_T=6 [mm]

i=1; η=15.7%

r_b=1 [mm]

e=0 [mm]

law: sin

y=x-sin(2πx)/(2π)

-initial spring deflection x_0=100 [mm]
-elastic constant of spring k=5 [N/mm]
-roll radius of the follower r_b=1 [mm] at a r_0=20 [mm].

Profile of rotary cam and translated lift with roll; Law of motion SIN-002;
Better efficiency η=17-70%;   Name: "4SIN-002"   ![n_{cam}=n/8]!

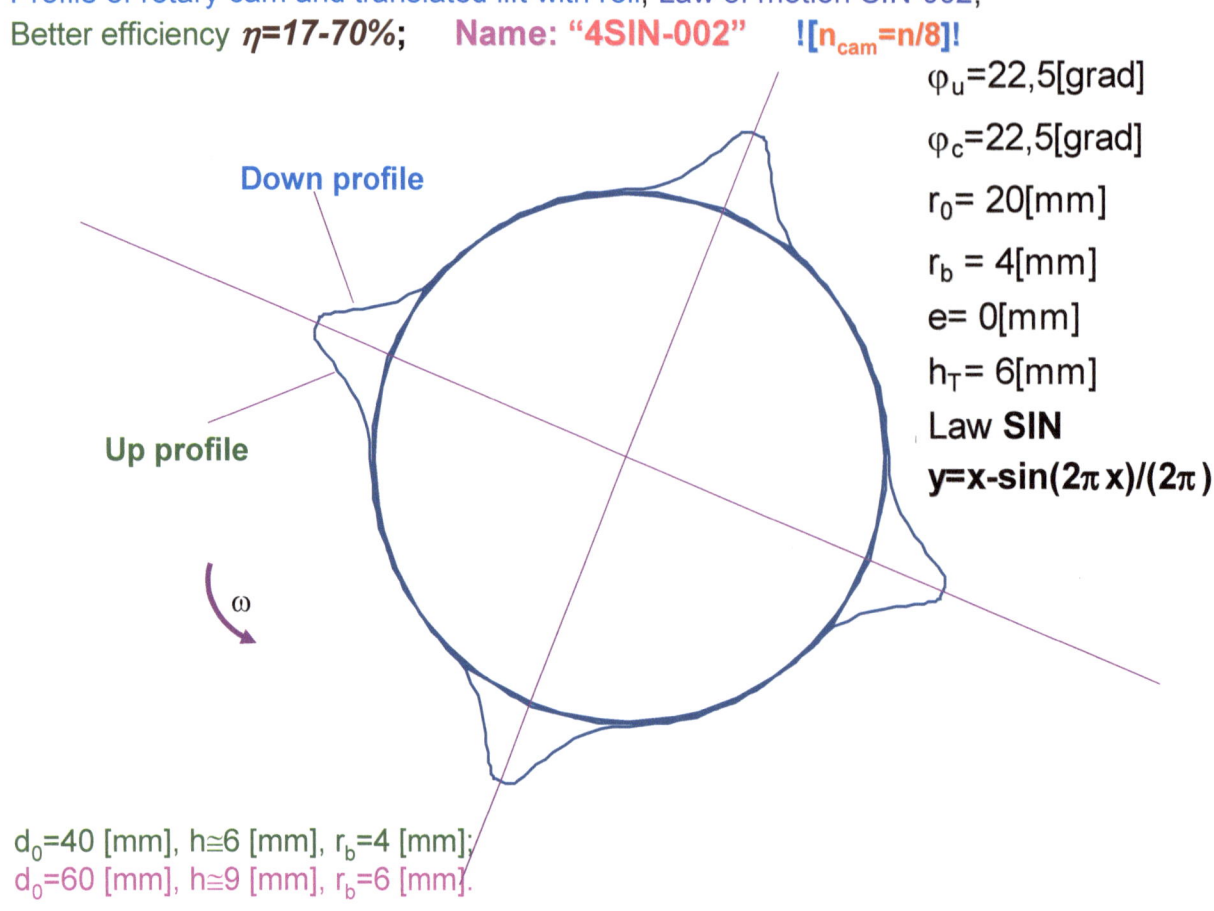

φ_u=22,5[grad]

φ_c=22,5[grad]

r_0= 20[mm]

r_b = 4[mm]

e= 0[mm]

h_T= 6[mm]

Law **SIN**

y=x-sin(2π x)/(2π)

Down profile

Up profile

ω

d_0=40 [mm], h≅6 [mm], r_b=4 [mm];
d_0=60 [mm], h≅9 [mm], r_b=6 [mm].

Dynamic analysis to the rotary cam and translated lift with roll; Law of motion SIN-002;
Better efficiency η=17-70%; Name: "4SIN-002"   ![n_{cam}=n/8]!

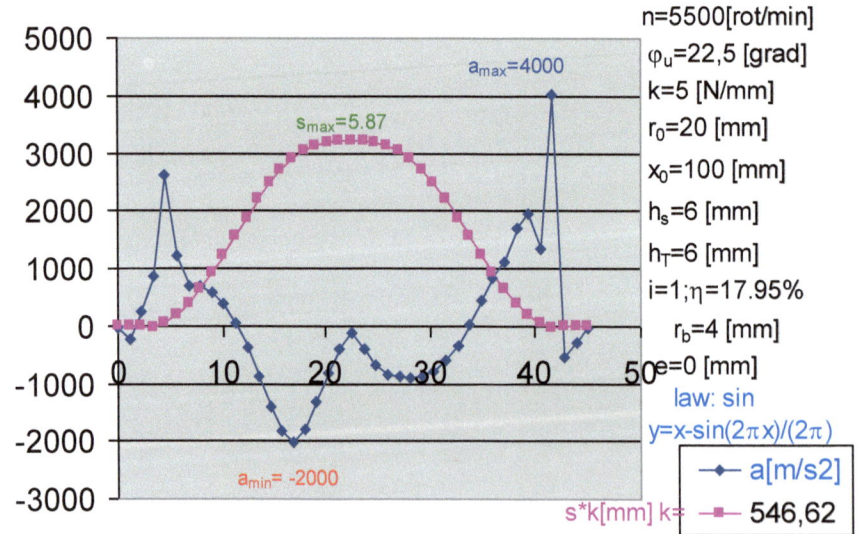

n=5500[rot/min]
φ_u=22,5 [grad]
k=5 [N/mm]
r_0=20 [mm]
x_0=100 [mm]
h_s=6 [mm]
h_T=6 [mm]
i=1;η=17.95%
r_b=4 [mm]
e=0 [mm]
law: sin
y=x-sin(2πx)/(2π)

a_{max}=4000
s_{max}=5.87
a_{min}= -2000
s*k[mm] k=

a[m/s2]
546,62

-initial spring deflection x_0=100 [mm]
-elastic constant of spring k=5 [N/mm]
-roll radius of the follower r_b=4 [mm] at a r_0=20 [mm].

Profile of rotary cam and translated lift with roll; Law of motion SIN-001;
Better efficiency **η=14-80%**;   Name: **"8SIN-001"**   **![n_{cam}=n/16]!**

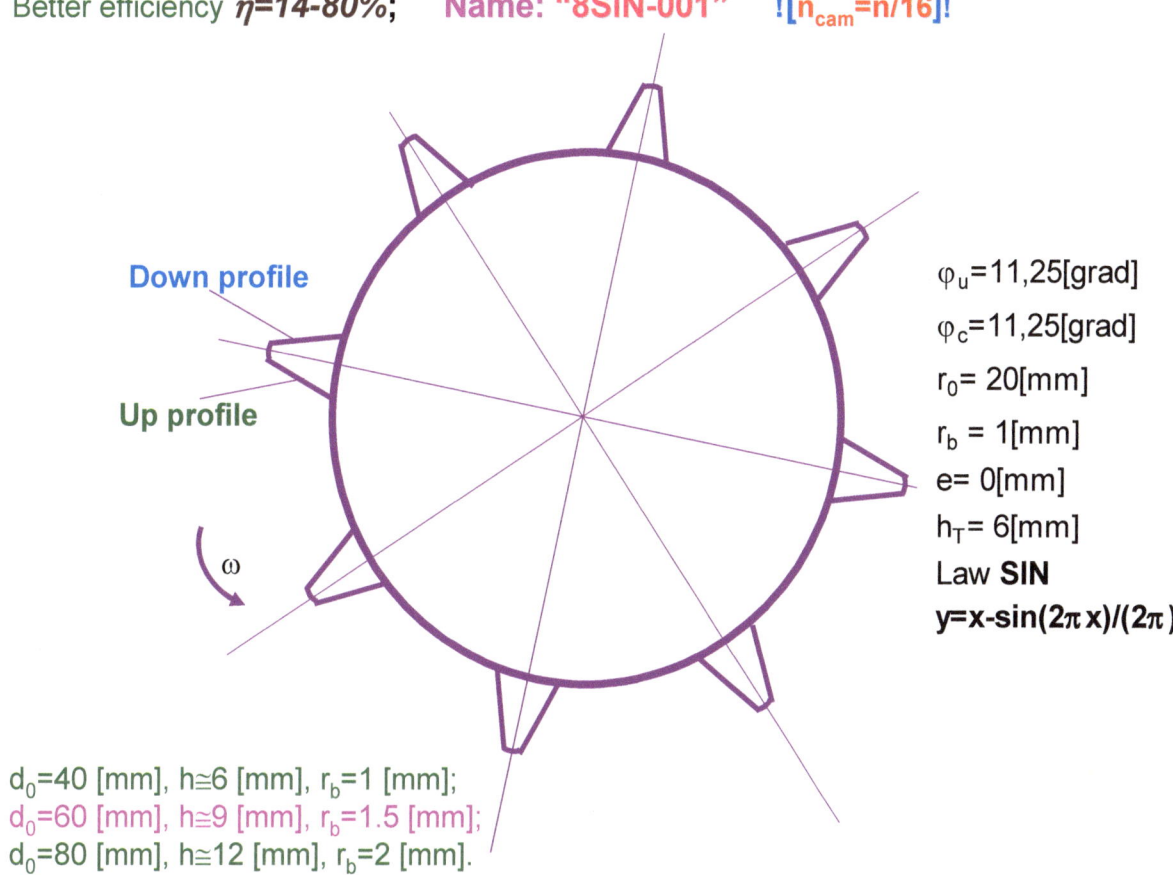

Down profile

Up profile

ω

φ_u=11,25[grad]

φ_c=11,25[grad]

r_0= 20[mm]

r_b = 1[mm]

e= 0[mm]

h_T= 6[mm]

Law **SIN**

y=x-sin(2πx)/(2π)

d_0=40 [mm], h≅6 [mm], r_b=1 [mm];
d_0=60 [mm], h≅9 [mm], r_b=1.5 [mm];
d_0=80 [mm], h≅12 [mm], r_b=2 [mm].

Dynamic analysis to the rotary cam and translated lift with roll; Law of motion SIN-001;
Better efficiency **η=14-80%**; Name: **"8SIN-001"**   **![n_{cam}=n/16]!**

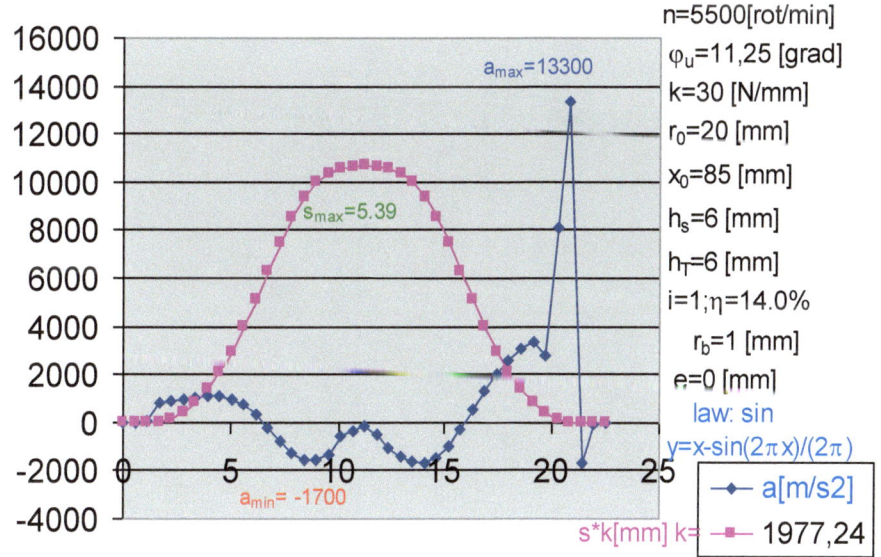

n=5500[rot/min]
φ_u=11,25 [grad]
k=30 [N/mm]
r_0=20 [mm]
x_0=85 [mm]
h_s=6 [mm]
h_T=6 [mm]
i=1;η=14.0%
r_b=1 [mm]
e=0 [mm]
law: sin
y=x-sin(2πx)/(2π)

a_{max}=13300

s_{max}=5.39

a_{min}= -1700

s*k[mm] k=

a[m/s2]
1977,24

-initial spring deflection x_0=85 [mm]
-elastic constant of spring k=30 [N/mm]
-roll radius of the follower r_b=1 [mm] at a r_0=20 [mm].

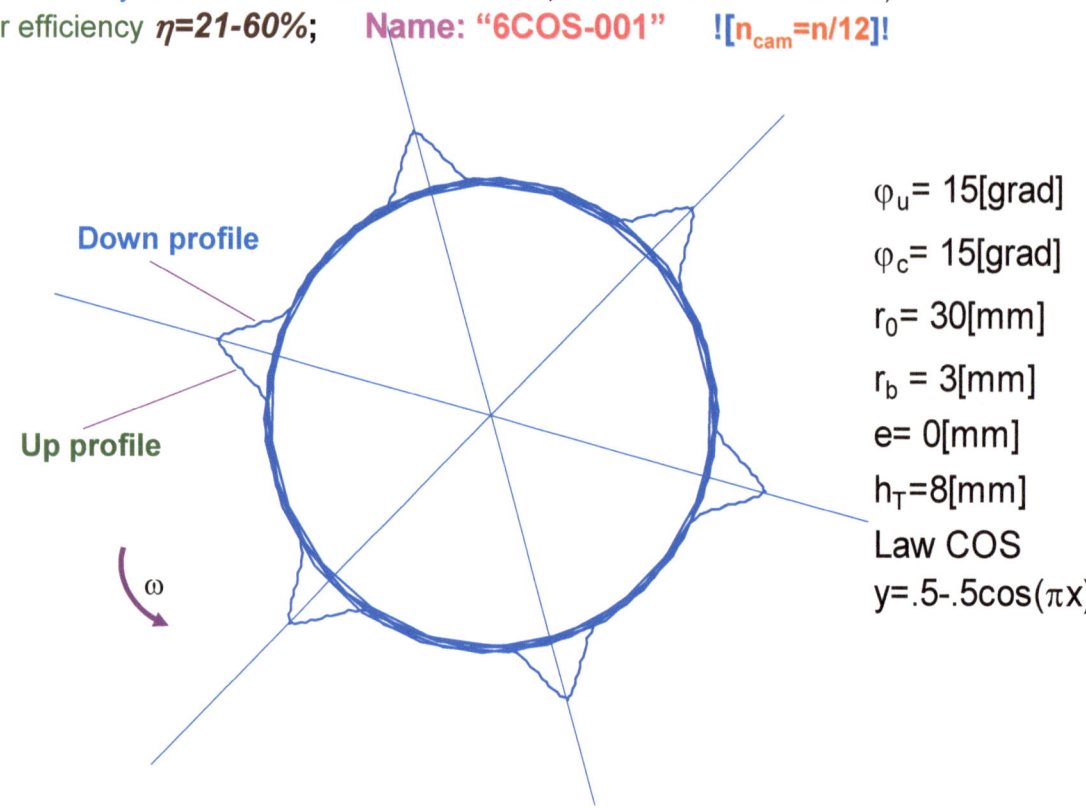

Down profile

Up profile

ω

φ_u= 15[grad]

φ_c= 15[grad]

r_0= 30[mm]

r_b = 3[mm]

e= 0[mm]

h_T=8[mm]

Law COS

$y = .5 - .5\cos(\pi x)$

d_0=60 [mm], h≅8 [mm], r_b=3 [mm];
d_0=90 [mm], h≅12 [mm], r_b=4.5 [mm].

Dynamic analysis to the rotary cam and translated lift with roll; Law of motion COS-001;
Better efficiency **η=21-60%**; Name: "6COS-001"   ![n_cam=n/12]!

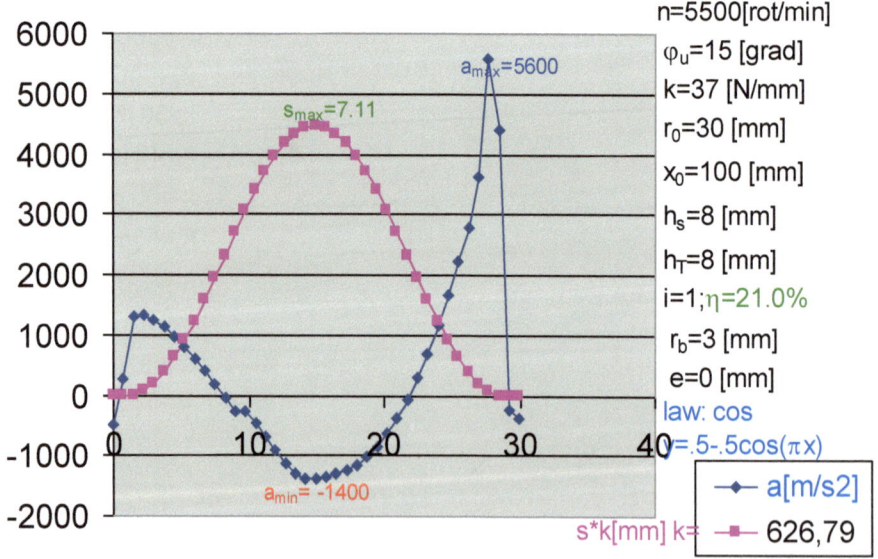

n=5500[rot/min]

φ_u=15 [grad]

k=37 [N/mm]

r_0=30 [mm]

x_0=100 [mm]

h_s=8 [mm]

h_T=8 [mm]

i=1;η=21.0%

r_b=3 [mm]

e=0 [mm]

law: cos

$y=.5-.5\cos(\pi x)$

a[m/s2]

s*k[mm] k= 626,79

a_{max}=5600

s_{max}=7.11

a_{min}= -1400

-initial spring deflection x_0=100 [mm]
-elastic constant of spring k=37 [N/mm]
-roll radius of the follower r_b=3 [mm] at a r_0=30 [mm].

Profile of rotary cam and translated lift with roll; Law of motion COS-001;
Best efficiency ***η=19-80%***; **Name: "12COS-001"** **![n_{cam}=n/24]!**

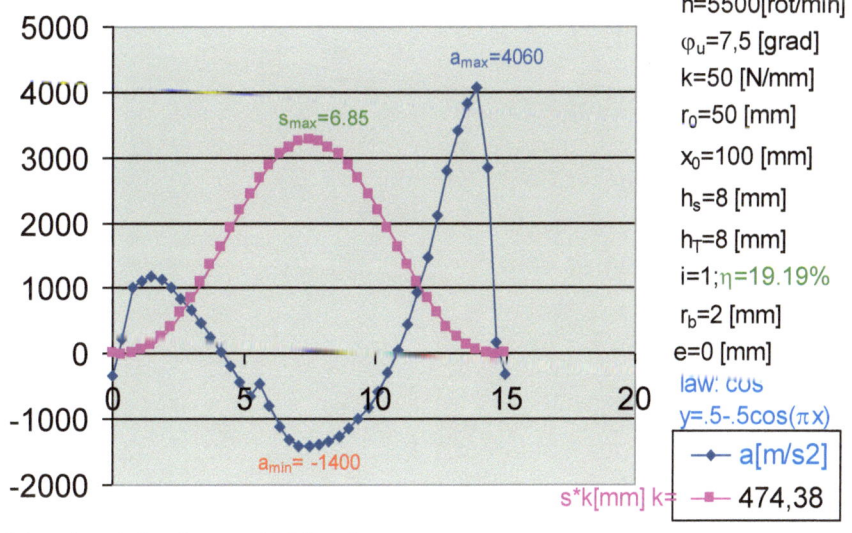

Down profile
Up profile
ω
φu= 7,5[grad]
φc= 7,5[grad]
r0= 50[mm]
rb = 2[mm]
e= 0[mm]
hT=8[mm]
Law COS
y=.5-.5cos(πx)
It would be desirable rounding the peaks!
d0=50 [mm], h≅4 [mm], rb=1 [mm].
d0=100 [mm], h≅8 [mm], rb=2 [mm];
d0=150 [mm], h≅12 [mm], rb=3 [mm].

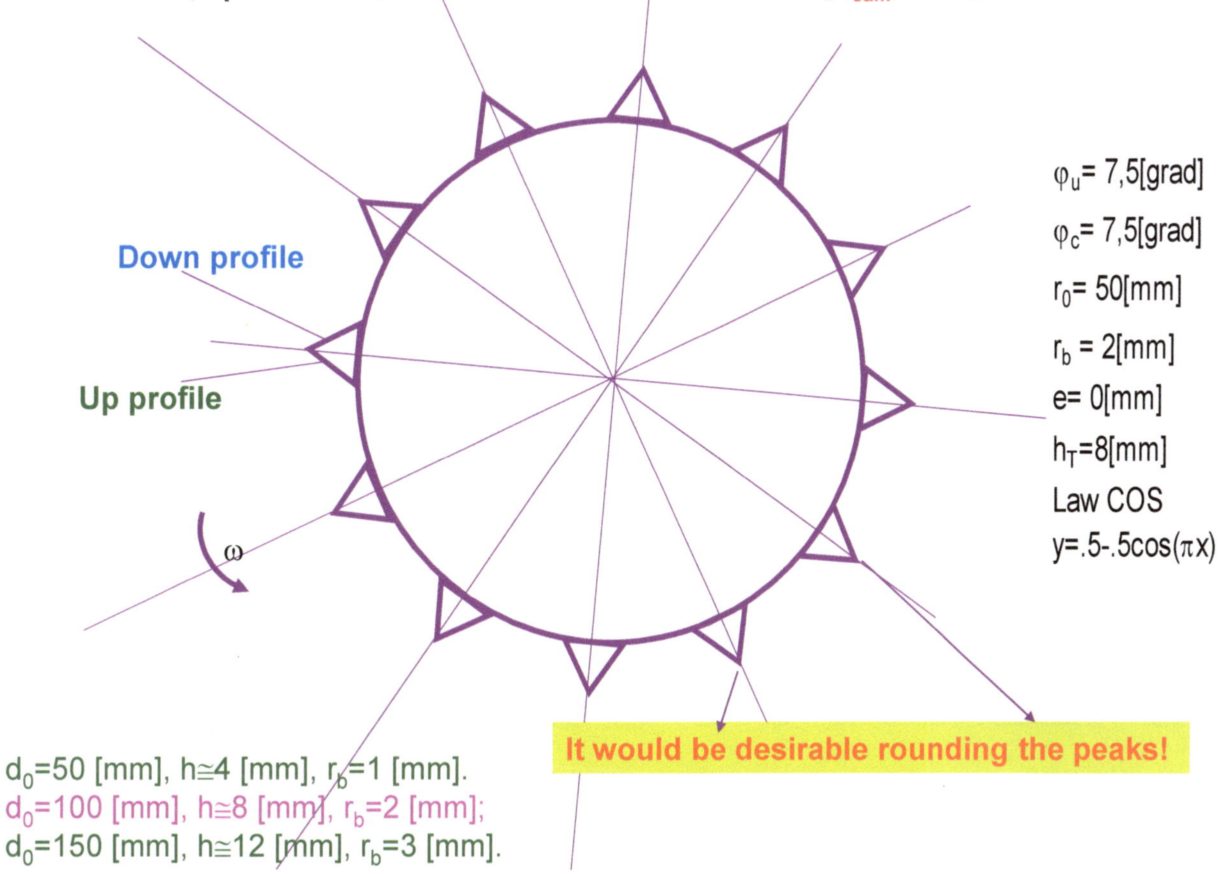

Dynamic analysis to the rotary cam and translated lift with roll; Law of motion COS-001;
Best efficiency ***η=19-80%***; **Name: "12COS-001"** **![n_{cam}=n/24]!**

5000
4000
3000
2000
1000
0
-1000
-2000
0
5
10
15
20
amax=4060
smax=6.85
amin= -1400
n=5500[rot/min]
φu=7,5 [grad]
k=50 [N/mm]
r0=50 [mm]
x0=100 [mm]
hs=8 [mm]
hT=8 [mm]
i=1;η=19.19%
rb=2 [mm]
e=0 [mm]
law: cos
y=.5-.5cos(πx)
a[m/s2]
s*k[mm] k= 474,38

-initial spring deflection x_0=100 [mm]
-elastic constant of spring k=50 [N/mm]
-roll radius of the follower r_b=2 [mm] at a r_0=50 [mm].

!All these matters are copyrighted!

See You Soon!